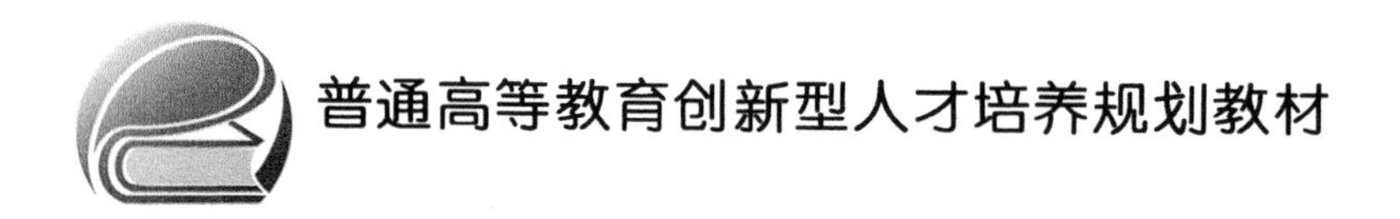

Basic of Mechanical Vibration

机械振动基础

Mao Qibo Li Yi

毛崎波 李奕 编著

北京航空航天大学出版社

Abstract

This book mainly describes vibration problems for linear discrete and continuous systems in engineering. This book provides the background and techniques for modeling, analysis, design, and control of vibration in mechanical engineering systems. This book is organized into 6 chapters, including introduction, free vibration of single-degree-of-freedom (SDOF) systems, harmonic excitation of SDOF systems, vibration of SDOF systems under general excitation, vibration of multiple-degree-of-freedom systems and vibration of continuous systems. Some background materials (such as mathematical background, basic of MATLAB, Laplace transform and technical terms) are presented in the appendices.

This book focuses on the linear vibration problems and emphasizes the ability to solve engineering vibration problems by using MATLAB software. Numerous examples and problems have been included both to assist the student in mastering the material and to demonstrate the applicability of the methods of analysis used in the book.

This book is suitable as a course textbook in a single-semester course for second-year or third-year undergraduate students, or for Master degree candidate in any branch of engineering such as aeronautical and aerospace, aircraft design, and mechanical engineering. The book can also serve as a valuable reference tool for practicing engineers with an interest in vibration problems.

图书在版编目(CIP)数据

机械振动基础 = Basic of Mechanical Vibration : 英文 / 毛崎波,李奕编著. -- 北京 : 北京航空航天大学出版社, 2019.12

ISBN 978-7-5124-3149-2

Ⅰ. ①机… Ⅱ. ①毛… ②李… Ⅲ. ①机械振动—高等学校—教材—英文 Ⅳ. ①TH113.1

中国版本图书馆 CIP 数据核字(2019)第 227292 号

Basic of Mechanical Vibration

机械振动基础

Mao Qibo Li Yi

毛崎波 李奕 编著

责任编辑 张冀青

*

北京航空航天大学出版社出版发行

北京市海淀区学院路 37 号(邮编 100191) http://www.buaapress.com.cn

发行部电话:(010)82317024 传真:(010)82328026

读者信箱: goodtextbook@126.com 邮购电话:(010)82316936

北京九州迅驰传媒文化有限公司印装 各地书店经销

*

开本:710×1 000 1/16 印张:15 字数:328 千字

2020 年 1 月第 1 版 2021 年 6 月第 2 次印刷 印数:1 001～1 500 册

ISBN 978-7-5124-3149-2 定价:49.00 元

PREFACE

With the recent interest in environmental issues, vibration problems have been a subject of intense interest for many years in modern industrial societies, such as maintaining high performance level and production efficiency. Various control techniques have been developed in different fields to reduce vibration levels. Although vibration control has been studied for long time, it remains and indeed becomes more challenging in many applications such as aerospace and aeronautic industry. Furthermore, it is important to understand and analyze the vibration performance of the engineering system when we try to control the structural and mechanical vibration levels.

This book provides the background and techniques for modeling, analysis, design, and control of vibration in mechanical engineering systems. This book is suitable as a course textbook in a single-semester course for second-year or third-year undergraduate students, or for Master degree candidate in any branch of engineering such as aeronautical and aerospace, aircraft design, and mechanical engineering. The book can also serve as a valuable reference tool for practicing engineers with an interest in vibration problems.

This book is an outgrowth of the first author's experience in teaching undergraduate and graduate courses in Basic of Mechanical Vibration, Active Noise and Vibration Control, for more than ten years.

The topics discussed in this book include free and forced vibration of the single-degree-of-freedom systems and multiple-degree-of-freedom systems. Vibrations of continuous systems are also presented. This book is organized around the concepts of vibration analysis of basic mechanical systems. Although the earlier chapters provide the basis for the later chapters, each chapter is written to be as self-contained as possible, with excerpts from earlier chapters provided as needed. This knowledge should be useful in the practice of vibration analysis or control, regardless of the application area or the branch of engineering.

Readers are expected to obtain skills ranging from the ability to perform insightful hand analysis to the ability to develop algorithms for numerical/computer analysis. A large number of examples are presented throughout the book so that the subject can be better understood. To fully master the analytical techniques, it is essential that the students solve many of the homework problems which are provided at the end of each chapter. In addition, many MATLAB programs are presented for additional study. Furthermore, the MATLAB GUI (Graphical User Interfaces) files were used to generate most of the results presented throughout the book. The reader is welcome to use them freely.

Mao Qibo wishes to appreciate the financial support from the National Natural Science Foundation of China (No. 51975266, No. 11464031), the Natural Science Foundation of Jiangxi, China (No. 20192BAB206024), the Six Talent Peaks Project of Jiangsu Province, China (No. 2017-KTHY-036) and the Publishing Foundation of Nanchang Hangkong University.

Moreover, we wish to thank the reviewers for their valuable comments and suggestions. We appreciate Ms. Ying Zeduan from Macau University of Science and Technology for her excellent work in improving the readability of the book. We gratefully acknowledge the editorial and production staff of Beihang University Press including Ms. Dong Rui.

In the end, we dedicate this book to our families.

Mao Qibo and Li Yi

August 2019

NOTES TO STUDENTS

Is it not a delight to acquire knowledge and put it into practice from time to time?

—Confucius (Chinese great educator)

Although mechanical engineering is an exciting and challenging discipline, you may be frightened during studying the mechanical vibration course. This book was written to try to prevent that. **A good textbook and a good professor are an advantage—but you are the one who does the learning.** If you keep the following ideas in mind, you will do very well to study mechanical vibrations:

- The basic of mechanical vibration provides a foundation for other courses in the mechanical engineering. For this reason, put in as much effort as you can. Study the course regularly.
- Mechanical vibration is a problem-solving subject, so we should learn through practice. Solve as many problems as you can. The best way to learn mechanical vibration is to solve a lot of problems by hand or by using computer software.
- MATLAB is a very useful software in signals analysis and other courses you may be taking. A brief tutorial on MATLAB is given in Appendix B, which help you to start learn this software. The best way to learn MATLAB is to try it once you know a few functions and commands.
- Attempt the review questions at the end of each chapter. They will help you discover some "tricks" not revealed in the class or in the textbook.

A short review of the mathematical formulas you may need is covered in Appendix A. Some terminology in English and in Chinese are listed in Appendix D.

Contents

Chapter 1 Introduction

1.1 Background

Vibration is a motion that repeats itself. Mechanical vibration is a repetitive, periodic, or oscillatory response of a mechanical system. This repetition may or may not perpetuate. The repetition also does not have to be a literal duplication. The rate of the vibration cycles is termed as "frequency". At low frequency ranges, these repetitive motions can be called as oscillations, because these motions are always clean and regular. However, it should be noted that any repetitive motion falls into the general class of vibration, which may have irregular and random behavior at high frequencies with low amplitudes. Nevertheless, the terms "vibration" and "oscillation" are often used interchangeably, as is done in this textbook.

Before the Industrial Revolution began, it was well-known that structures were usually of very high quality because the heavy timbers and stonework were used in their fabrication. So the dynamic responses of these structures were extremely low due to the large mass and low amplitude of the excitation sources. Furthermore, a structure was usually a combination of parts fastened together to create a supporting framework at that time. These constructional methods usually produced a structure with very high inherent damping, which also gave a low structural response to dynamic excitation.

However, from the last 200 years, with the advent of relatively strong lightweight materials such as cast iron, steel, plastic and aluminium, the mass of structures can be reduced to fulfill some particular functions. The structural mass can be further reduced through advanced additive manufacturing technology and by using new materials. At the same time, the amplitudes of the vibration exciting sources have increased quickly, because the engines have been more powerful and higher rotational speeds.

In the present day, very few structures with acceptable dynamic performance can be designed without the necessary vibration analysis, because the structural mass is more and more reduced and the exciting forces are increasing.

Generally speaking, we hope to reduce or eliminate the vibration in most machines, structures and dynamic systems, because the vibration will cause unpleasant motions and structural acoustical noise. And the vibration may lead to fatigue or failure of the structure and machine. The structural vibration will also cause the energy losses. Therefore, it is essential to perform a vibration analysis of any proposed structures.

It is well-known that there have been very large cases of systems failing or not meeting performance targets because of resonance or excessive vibration of one or several components. Because unwanted vibrations have the very serious effects on dynamic systems or structures, it is essential to carry out vibration analysis as an inherent part of their design. Although eliminating structural vibration in most cases is not practically possible. However, any necessary modifications can most easily be made to reduce the structural vibration as low as possible.

It is usually much easier to analyze and modify a structure at the design stage than it is to modify a structure with undesirable vibration characteristics after it has been built. However, it is sometimes necessary to be able to reduce the vibration of existing structures due to inadequate initial design. For example, if the function of the structure and/or the environmental conditions is changed, the system/structures may arise the undesirable vibration at some frequencies. So techniques for the analysis of structural vibration should be applicable to existing structures as well as to those in the design stage. The solution to vibration problems may be different depending on whether the structure exists.

In this textbook, practical applications and design considerations related to modifying the vibrational behavior of mechanical systems and structures will be studied. This knowledge should be useful in the practice of vibration regardless of the application area or the branch of engineering, for example, in the analysis, design, construction, operation, and maintenance of complex structures such as the Space Shuttle and the International Space Station. The long and flexible components, which would be prone to complex "modes" of vibration, are present. The structural design should take this into consideration. Also, functional and servicing devices such as robotic manipulators (such as Canadarm shown in Fig. 1. 1) can

Fig. 1. 1 The Canadarm

give rise to vibration interactions that need to be controlled for accurate performance. The approach used in the textbook is to introduce practical applications of vibration in the very beginning, and then integrate these applications and design considerations into fundamentals and analytical methods throughout the text.

In summary, today's structures often contain high-energy sources that create strong vibration excitation problems, and the structures of modern constructions often have low mass and low inherent damping. Therefore careful design and analysis should be performed to avoid resonance or the undesirable structural vibration performance.

1.2 Study of Vibration

It should be noticed that most human activities also involve vibration in one form or others. For example, we hear because our eardrums vibrate. Breathing is associated with the vibration of lungs and speech requires the vibration motion of larynges and tongues. Human walking involves oscillatory motion of legs. In fact, without vibrations, we could not recognize the universe around us at all!

Early scholars and researchers in the area of vibration have developed many mathematical theories which can be used to describe the different vibration phenomena (such as linear or nonlinear vibration). Recently, researchers more and more concentrate on the vibration problems in the engineering applications. For example, the design of low noise and vibration machines and foundations, how to reduce and control engines and turbines vibrations. Although the vibration problems have been studied for long time, it remains and indeed becomes more challenging in many applications such as aerospace and aeronautic industry. Notice that the vibration problems are ubiquitous in engineering. It means that the study of vibrations is extremely important.

Most vehicles have vibrational problems due to the inherent unbalance in the engines (internal combustion engines, turbojet engines and so on.). The unbalance may be due to faulty design or poor manufacture. However, sometimes the unbalance vibration is inevitable, such as for washing machines. The unbalance vibration can be dangerous, for example, the wheels of some locomotives can rise more than a centimeter off the track at high speeds due to imbalance. The bearing and gear failures in turbines usually are caused by vibration. Naturally, the structures designed to support heavy centrifugal machines (such as motors and turbines) or reciprocating machines (such as steam, gas engines and reciprocating pumps) are also subjected to vibration. In all these situations, the structure or machine component subjected to vibration can fail because of material fatigue resulting from the cyclic variation of the induced stress. Furthermore, the structural vibration also creates excessive noise. Vibrations from railways, heavy transport or construction work are a nuisance to many people and may disturb precision instruments. In machines, vibration can loosen fastening pieces (such as screws and bolts). In metal cutting processes, the lathe vibration can lead to a poor surface finish due to the chatter phenomenon. When the frequency of the external excitation coincides with the natural frequency of vibration of a machine or structure, a phenomenon known as resonance could happens. The resonance will lead to large amplitude vibration and may cause system failure. The literature is full of accounts of system failures brought about by resonance and excessive vibration of components and systems, as shown in Fig. 1. 2. To reduce the vibration on machines and structures, vibration analysis and testing has become a standard procedure in the design and development of most engineering structures and systems (see Fig. 1. 3).

Fig. 1. 2 Tacoma Narrows bridge during wind-induced vibration
(The bridge opened on July 1, 1940, and collapsed on November 7, 1940)

Fig. 1. 3 Aircraft ground vibration testing

Notice that much of the noise is created by the vibration of solid, elastic bodies. This type of noise is referred to as structure-borne sound. In many engineering systems, humans can be seen as an integral part of the system. The structure-borne sound due to vibration can lead to discomfort, low efficiency or loss of property. For example, the vibration and noise generated by engines cause trouble to people. Low levels of vibration mean reduced noise and an improved work environment. And sometimes vibration may damage precision instruments. Waves and wind cause vibrations that have negative effects on the safety and service life of offshore platforms and wind turbines. Therefore one of the important issues of vibration study is to control or reduce vibration of the mechanical structures. Recently, vibration control has drawn more and more intensive efforts from mechanical and structural engineer.

It should be noticed that the vibration can be useful in many industrial applications, in spite of its harmful effects. For example, the loudspeaker can produce sound due to the vibration of the coil and diaphragm. String instruments (such as violin and guitar) produce sound from vibrating strings. In fact, the applications of vibratory equipment have increased considerably in recent years. For example, a vibrating sieve separates particles of fine sand and coarse sand; the vibrating mode in our mobile phone to alert us without disturbing others. And vibration is put to work in washing machines, vibratory conveyors and electric toothbrushes, etc. Vibration is also used in pile driving, vibratory testing of materials, vibratory finishing processes, earthquake monitoring (seismograph) (see Figs. 1.4 and 1.5).

Fig. 1.4　The pile driving and vibratory finishing (oscillating mill)

Fig. 1.5　The useful vibration

1.3 Basic Concepts of Vibration

Any motion that repeats itself after an interval of time is called vibration or oscillation. The rocking of a pendulum and the motion of a plucked string are typical examples of vibration. The theory of vibration deals with the study of oscillatory motions of bodies and the forces associated with them.

Generally speaking, a vibration system includes three parts, that is (1) a means for storing potential energy (spring or elasticity), (2) a means for storing kinetic energy (mass or inertia), and (3) a means by which energy is gradually lost (damper).

The vibration of a system involves the transformation of its potential energy to kinetic energy and of kinetic energy to potential energy, alternately. If the system is damped, some energy is dissipated in each cycle of vibration. If a state of steady vibration wants to be maintained, the damped system must be excited by an external source.

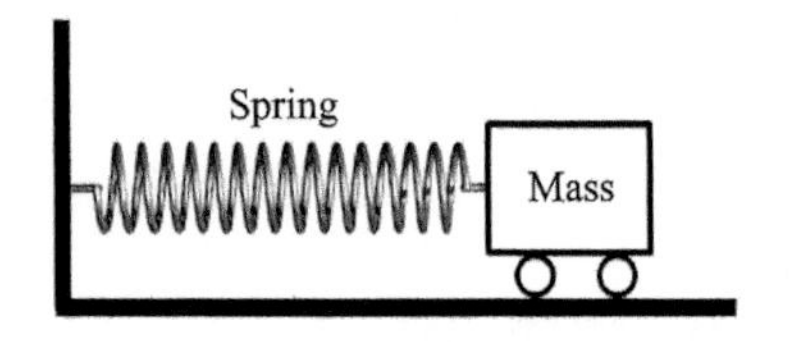

Fig. 1.6 A spring-mass system

A mass connected to a horizontal spring is a typical vibration system, as shown in Fig. 1.6. The mass is the component responsible for kinetic energy, and the spring is that for potential energy.

As another example, consider the vibration of the simple pendulum, as shown in Fig. 1.7. Notice that the pendulum does not have a spring-like component for potential energy. In fact, the mass plays a dual role for both kinetic energy and potential energy. This is an example to show that the mass and stiffness components in a vibration system are NOT necessarily the separate components.

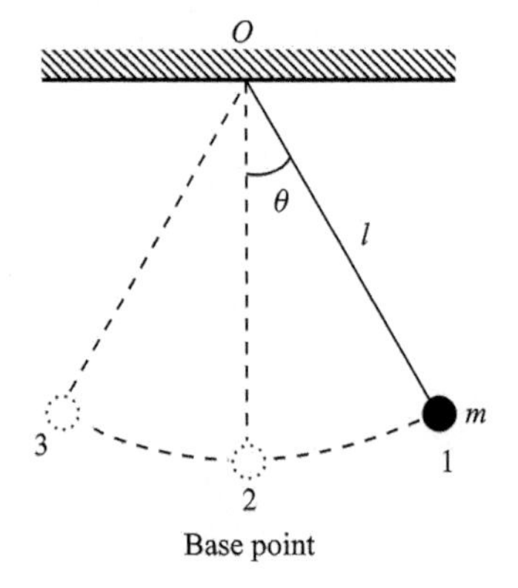

Fig. 1.7 A simple pendulum

As shown in Fig. 1.7, if the mass m is released after being given an angular displacement θ. At location 1, both of the velocity and kinetic energy of the mass are zero. But it has a potential energy due to the amplitude with respect to the base location 2. Since the gravitational force mg induces a torque about the point O, the mass starts swinging to the left from location 1. This gives the mass certain angular acceleration in the clockwise direction, and by the time it reaches the base location 2, all of its potential energy will be transferred into kinetic energy. Clearly, the mass will continue to swing to location 3 due to the angular acceleration. However, as it passes the base location 2, a counterclockwise torque due to gravity starts acting on the mass, and the angular acceleration causes the mass to decelerate. The velocity of the mass reduces to zero at the left extreme location 3. At this time, all the kinetic energy of the mass will be converted to potential energy. Again due to the gravity torque, the mass continues to obtain a counterclockwise angular velocity. Hence the mass starts swinging back with gradually increasing velocity and passes the base location (position 2) again. This process keeps repeating, and the pendulum will have oscillatory motion. However, in practice, the magnitude of oscillation will gradually decrease and the pendulum ultimately stops due to the resistance (damping) offered by the surrounding medium (air). It means that some energy is dissipated in each cycle of vibration due to damping by the air.

1.3.1 Degrees of freedom

Definition: The minimum number of independent coordinates required to determine completely the positions of all parts of a system at any instant of time defines the degree of freedom of the system.

To be able to study the vibration of a dynamic system, it isimportant to know how many degrees of freedom the system has before the analysis begins. The number of the degrees of freedom of a vibration system is the minimum number of independent coordinates required to determine completely the motion of all parts of the system at any instant of time.

For example, a mass-spring-damper system shown in Fig. 1.8 has two masses. It is clearly that each mass is able to move independently, because the mass can move physically if another mass is fixed at any location. Therefore, there are two degrees of freedom for this system. The degrees of freedom can be represented by different coordinates. For the mass-spring-damper system shown in Fig. 1.8, the two coordinates used for analysis can be either x_1 or x_2, or they can be x_1 and (x_1-x_2), or even other choices. A different selection of coordinates will lead to different motion equations. However, since we are dealing with the same system, we should expect that the

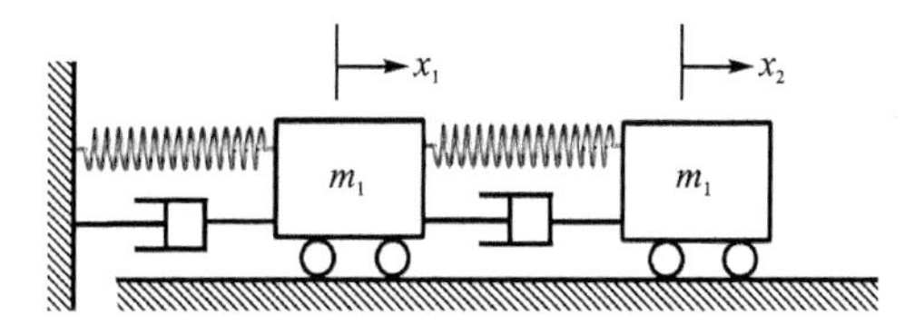

Fig. 1.8 A two-degree-of-freedom system

natural frequencies derived from these different equations are the same, regardless the choice of coordinates.

1.3.2 Newton's laws

Newton's laws will be used throughout this textbook. The concept of Newton's laws is briefly given as follows.

Newton's first law: If there are no forces acting upon a particle, then the particle will move in a straight line with constant velocity.

Newton's second law: A particle acted upon by a force moves so that the force vector is equal to the time rate of change of the linear momentum vector.

Newton's third law: When two particles exert forces upon one another, the forces lie along the line joining the particles and the corresponding force vectors are the negative of each other.

1.3.3 Vibration classification

There are different ways of classifying the types of vibration. These classifications may overlap. Table 1.1 summarizes these vibrations and their brief descriptions.

Table 1.1 Different types of vibration

Reference terms	Vibration types	Descriptions
External excitation	Free vibration	Vibration induced by initial conditions only
	Forced vibration	Vibration subjected to one or more continuous external inputs
Presence of damping	Undamped vibration	Vibration with no energy loss or dissipation
	Damped vibration	Vibration with energy loss
Linearity of vibration	Linear vibration	Vibration for which superposition principle holds
	Nonlinear vibration	Vibration that violates superposition principle
Predictability	Deterministic vibration	The value of vibration is known at any given time
	Random vibration	The value of vibration is not known at any given time but the statistical properties of vibration are known
Number of degree of freedom	Discrete	Systems with a finite number of degrees of freedom
	Continuous	Systems involving continuous elastic members, have an infinite number of degrees of freedom

1.4 Organization of the Book

This textbook provides the background and techniques for modeling and analysis of vibration in engineering structures and systems. This knowledge will be useful in the practice of vibration, regardless of the application area or the branch of engineering. MATLAB examples are given throughout the text. Readers are expected to obtain skills ranging from the ability to perform insightful hand analysis to the ability to develop algorithms for numerical analysis.

The textbook consists of 6 chapters and 4 appendices. Many examples and problems are included. Some background materials (such as mathematical background, basic of MATLAB, Laplace transform and technical term in Chinese) are presented in the appendices, rather than in the main text, in order to avoid interference with the continuity of the subject matter.

The present introductory chapter provides some background materials on the subject of vibration engineering, and sets the course for the study. It gives the objectives and motivation of the study and indicates key application areas.

Chapter 2 provides the basics of free vibration analysis of Single-Degree-Of-Freedom (SDOF) systems. Both undamped and damped systems are studied. An energy-based approximation of a distributed parameter system (a heavy spring) to a lumped-parameter system is developed in detail. The logarithmic decrement method of damping measurement is presented. Although this chapter primarily considers SDOF systems, the underlying concepts can be easily extended to Multiple-Degree-Of-Freedom (MDOF) systems.

Chapter 3 concerns the SDOF system under harmonic excitation. First, the response of a vibrating system to harmonic (sinusoidal) excitation forces (inputs) is analyzed. Specifically, force transmissibility and motion transmissibility are studied and their complementary relationships are highlighted. The force transmissibility and motion transmissibility used in the practice of vibration, particularly in vibration isolation, are discussed.

Chapter 4 presents the solutions for the SDOF systems under general force excitation. Various type forces in vibration engineering are discussed. The technique of Fourier series, convolution integral and Laplace transform analysis is introduced and linked to the concepts presented in this chapter.

Chapter 5 deals with the vibration analysis for Multi-Degree-Of-Freedom (MDOF) systems. The total number of possible independent, incremental motions of these inertia elements is the number of degrees of freedom of the system. The representation of a general MDOF vibrating system by a differential equation model is given, and methods of obtaining such a model are discussed. Several approach, such as the Newtonian approach, the influence coefficient approach and Lagrange's equation are given for determining the mass and stiffness matrices. The concepts of natural frequencies and mode shapes are discussed, and the procedure for determining these characteristic quantities is developed based on MATLAB. The responses of the damped MDOF systems are also studied.

Chapter 6 studies continuous vibrating systems, such as strings, beams and plates. The influence of system boundary conditions on the modal problem is presented.

Questions

1.1 What is meant by vibrations?

1.2 What is force vibration?

1.3 What is Degrees Of Freedom (DOF) in vibration systems?

Chapter 2 Free Vibration of Single-Degree-Of-Freedom(SDOF) Systems

For many dynamic systems, the relationship between restoring force and deflection is approximately linear for small deviations about some reference. If the system is complex (such as, an aircraft wing that requires numerous variables to describe its properties), it is possible to transform it into a number of SDOF linear vibration problems by using the normal modes of the system. And the motion of a SDOF system can be described using a single scalar second-order ordinary differential equation. The SDOF problem is also fundamental to understanding the principles of the vibration structures.

2.1 The Basic Mechanical Components

The theory of vibration deals with the dynamic response of vibration systems to different excitations. Since vibration can be regarded as the transformation between the kinetic energy and potential energy, a vibration system has to include components of storing (and releasing) both energies. The former is often done by a mass and the latter by a spring. Analytical models of vibration systems usually consist of masses, springs and dampers. They are the basic mechanical components of the system. In theory, relationship between the force, displacement and its time derivatives, velocity and acceleration, are presented as follows.

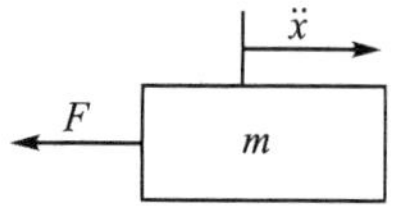

The inertial force is equal to the product of the mass m and its acceleration $\ddot{x}$ with opposite direction of the acceleration, as shown in Fig. 2.1. We have

Fig. 2.1 The force due to mass

$$F = -m\ddot{x} \tag{2.1}$$

The restoring spring force is proportional to the product of the spring stiffness k and the relative displacement of its ends, as shown in Fig. 2.2(a), we have

$$F = -k(x_1 - x_2) \tag{2.2a}$$

where the minus sign indicates that the restoring spring force opposes the relative displacement.

For a string which is fixed at one end, as shown in Fig. 2.2(b), the restoring spring force can be simplified as

$$F = -kx \tag{2.2b}$$

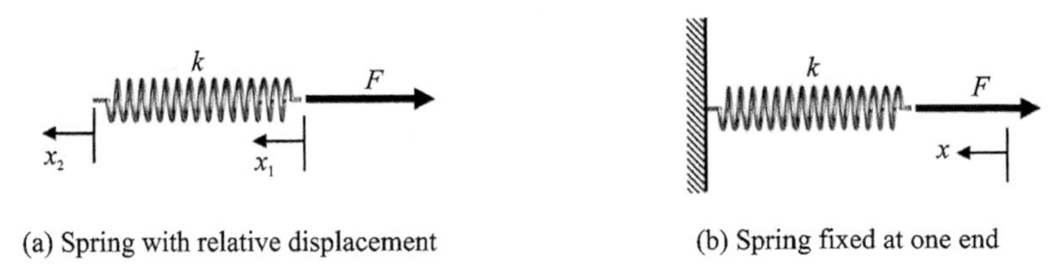

(a) Spring with relative displacement (b) Spring fixed at one end

Fig. 2.2 The force due to spring under different stiffnesses

The damping force due to a viscous damper is proportional to the product of the damper constant c and the relative velocity of the damper's ends, as shown in Fig. 2.3(a), we have

$$F=-c(\dot{x}_1-\dot{x}_2) \tag{2.3a}$$

where the minus sign indicates that the damping force resists an increase in velocity.

Similar, if the damper is fixed at one end, as shown in Fig. 2.3(b), then we have

$$F=-c\dot{x} \tag{2.3b}$$

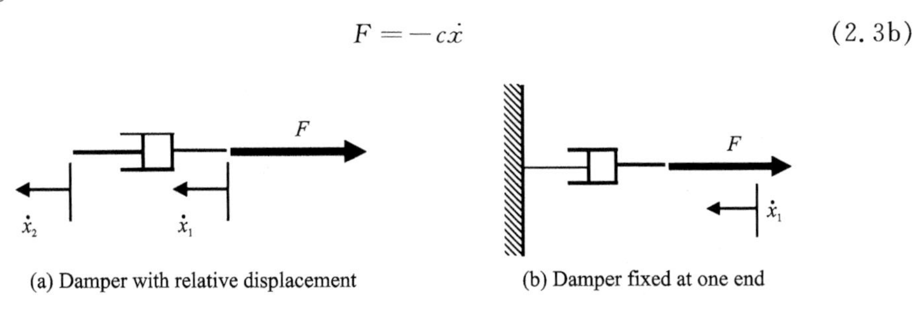

(a) Damper with relative displacement (b) Damper fixed at one end

Fig. 2.3 The force due to different dampers

2.2 Free Vibration of Undamped Systems

2.2.1 Modelling of undamped SDOF systems

The generic model for a SDOF system is a mass connected to a linear spring and a linear viscous damper (termed as a mass-spring-damper system). Because of its mathematical form, the mass-spring-damper system will be used as the baseline for analysis of the SDOF system. In particular, the differential equation of motion will be derived for the mass-spring-damper system. It will then be shown that the time response of this system is the sum of the **zero input response** and the **zero initial condition response.** In this chapter we will focus our attention on the zero input response, i. e., the response of the system to a given set of initial conditions. Several examples of SDOF systems will be presented. In each of these examples the differential equation will be derived and will be shown to have the same mathematical form as the generic mass-spring-damper system.

A vibration system containing the masses and springs only is called an undamped system. It means that there is NO damper (or the values of damper are equal to zero) in the undamped system. Assume that a mass m is suspended at the end of a spring, as shown in Fig. 2.4. The weight of the mass stretches the spring by a length Δ to reach a static state (the equilibrium position of the system).

From Fig. 2.4, it can be found that there are two coordinates can be used for analysis (either y or x). First, the analysis based on y-coordinate is presented. Clearly, $y(t)$ is the displacement, as a function of time, of the mass relative to its undeformed position. Assume that downward is positive. Therefore, $y > 0$ means that the spring is stretched, while $y < 0$ means that the spring is compressed. The mass is then assumed to be set in vibration.

The distance Δ is the static deflection of the system. For the system, as shown in Fig. 2.4, at equilibrium position, it is easy to obtain

$$mg = k\Delta \tag{2.4}$$

Based on the free body diagram, the equation of motion for Fig. 2.4 based on y-coordinate (coordinate origin at the undeformed position) can be expressed as

$$m\ddot{y} + ky = mg \tag{2.5}$$

Notice that there is a static force term on the right of Eq. (2.5). However, if we assume that the coordinate origin at the equilibrium position, it means that the motion equation is based on x-coordinate. Thus we have

$$y = x + \Delta \tag{2.6}$$

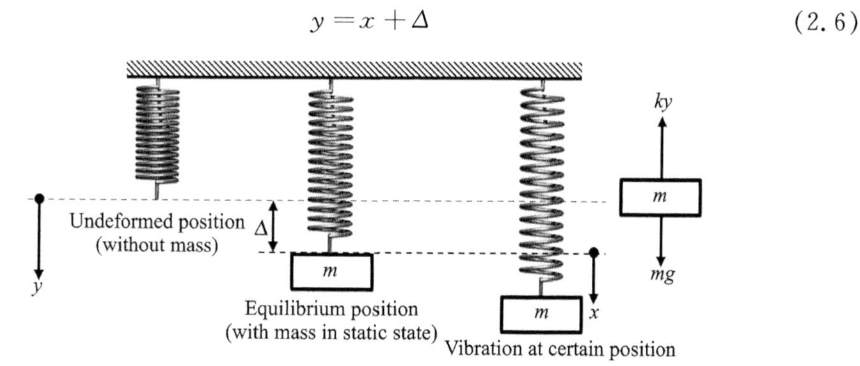

Fig. 2.4 Free vibration of an undamped mass-spring system

Substituting Eq. (2.4) into Eq. (2.6), we get

$$y = x + \frac{mg}{k} \tag{2.7}$$

Since m, g and k are constants, from Eq. (2.7), it is easy to obtain

$$\ddot{y}=\ddot{x} \tag{2.8}$$

Substituting Eqs. (2.7) and (2.8) into Eq. (2.5), we obtain the equation of motion which is measured from the static equilibrium position, yields

$$m\ddot{x}+kx=0 \tag{2.9}$$

Comment: It is always simpler to express the vibration motion equation by using a coordinate system with origin at the static equilibrium position, because the equation of motion expressed with reference to the static equilibrium position of the dynamic system will not be affected by gravity forces. Although example in Fig. 2.4 (the free vibration of an undamped mass-spring system) is a specific case, this conclusion holds in general. For this reason, displacements in all future discussions will be referenced from the static equilibrium position.

After we obtain the motion equation, next step is to determine the solution of Eq. (2.9). Notice that Eq. (2.9) is a linear, second-order, ordinary differential equation with constant coefficients. The general solution of Eq. (2.9) can be expressed as

$$x(t)=A\sin(\omega_n t)+B\cos(\omega_n t) \tag{2.10}$$

where A and B are unknown coefficients to be determined by the initial conditions, the natural frequency ω_n can be expressed as $\omega_n=\sqrt{\frac{k}{m}}$.

Proof:

Substituting Eq. (2.10) into Eq. (2.9), we have

$$m\ddot{x}+kx=-m\omega_n^2\left[A\sin(\omega_n t)+B\cos(\omega_n t)\right]+k\left[A\sin(\omega_n t)+B\cos(\omega_n t)\right] \tag{2.11}$$

Notice that $\omega_n=\sqrt{\frac{k}{m}}$, it means that Eq. (2.11) can be rewritten as

$$m\ddot{x}+kx=-m\frac{k}{m}\left[A\sin(\omega_n t)+B\cos(\omega_n t)\right]+k\left[A\sin(\omega_n t)+B\cos(\omega_n t)\right]=0 \tag{2.12}$$

From above analysis, it can be found that the solution in Eq. (2.10) satisfies the differential motion equation given in Eq. (2.9). Furthermore, according to Eq. (2.10), it can be found that the solution of Eq. (2.9) can be expressed in terms of a linear combination of two independent functions.

It should be noted that the natural frequency ω_n is the frequency at which the system tends to oscillate in the absence of any damping. A motion of this type is called simple harmonic motion. It is sinusoidal in time and demonstrates a single natural frequency. And A and B are arbitrary real constants which can be determined by the initial conditions.

Comment: Just like everywhere else in calculation, the angle is measured in radians, and the (angular) frequency is given in radians per second. The frequency is NOT given in hertz (which measures the number of cycles or revolutions per second). Instead, their relation is 2π rad/s=1 Hz.

Period is the time taken for one vibration cycle. Its symbol is T and its unit is seconds. Therefore, the (natural) period of the oscillation can be written as

$$T=\frac{2\pi}{\omega_n}(\text{seconds})$$

2.2.2 Simple harmonic motion

As shown in Fig. 2. 4, if the mass is displaced slightly from equilibrium position, it will oscillate about its equilibrium position. This type vibration (free vibration of undamped SDOF system) can be called as Simple Harmonic Motion (SHM). To show a more clear picture of how the solution in Eq. (2. 10) behaves, we can simplify Eq. (2. 10) with trigonometric function and rewrite it as

$$x(t)=R\sin(\omega_n t+\delta) \tag{2.13}$$

where $R=\sqrt{A^2+B^2}$, is the amplitude of the displacement. The angle δ is termed as the phase or phase angle of displacement. It indicates how much $x(t)$ lags (when $\delta>0$), or leads (when $\delta<0$) relative to $\sin(\omega_n t)$. The phase angle satisfies the following relation, that is $\tan\delta=\frac{B}{A}$.

More explicitly, it can be calculated by

$$\delta=\arctan\left(\frac{B}{A}\right),\ \text{if } A>0 \tag{2.14a}$$

$$\delta=\arctan\left(\frac{B}{A}\right)+\pi,\ \text{if } A<0 \tag{2.14b}$$

$$\delta=\frac{\pi}{2},\ \text{if } A=0 \text{ and } B>0 \tag{2.14c}$$

$$\delta=-\frac{\pi}{2},\ \text{if } A=0 \text{ and } B<0 \tag{2.14d}$$

The angle is undefined if $A=B=0$.

From Eq. (2. 13), it is clearly that the motion of the mass in Fig. 2. 4 is sinusoidal in time and demonstrates a single natural (or resonant) frequency ω_n.

2.2.3 MATLAB examples

An example of simple harmonic motion can be plotted by MATLAB. For example, there are functions $x(t)=\sin 2\pi t+\cos 2\pi t$ and $y(t)=\sin 2\pi t-\cos 2\pi t$. To plot these functions, the following MATLAB codes can be used to draw the results, as shown in Fig. 2. 5. From Fig. 2. 5, it can be found that the phase angles δ for $x(t)=\sin 2\pi t+\cos 2\pi t$ and $y(t)=\sin 2\pi t-\cos 2\pi t$ are $\pi/4$ and $-\pi/4$, respectively.

Line	MATLAB Codes	Comments
1	clear all, close all	Remove all variables in workspace and close all figures. It is a good way to start a new MATLAB script (not function)
2	t = linspace(0, 3, 1e3);	The linspace function generates linearly spaced vectors. It generates a row vector t of 1 000 points linearly spaced form 0 to 3
3	x = sin(2 * pi * t) + cos(2 * pi * t);	Calculate $x(t) = \sin 2\pi t + \cos 2\pi t$
4	y = sin(2 * pi * t) − cos(2 * pi * t);	Calculate $y(t) = \sin 2\pi t - \cos 2\pi t$
5	figure(1)	Create a new figure
6	plot(t, x,'k', 'linewidth', 2)	Plot the x by solid line (black color and linewidth=2 points)
7	hold on	Retain the current plot and certain axes properties so that subsequent graphing commands add to the existing graph, it is quite useful for multiple-line plot
8	plot(t, y,'b--', 'linewidth', 2)	Plot the y by dashed line (blue color and linewidth=2 points)
9	grid on	To displays the major grid lines for the current axes
10	xlabel ('{\itt} (Second)', 'fontsize', 14, 'fontname', 'times new roman')	Label the x-axis. "\it" is for italic font
11	ylabel ('{\itx}', 'fontsize', 14, 'fontname','times new roman')	Label the y-axis
12	legend('sin(2\pit) + cos(2\pit)', 'sin(2\pit) - cos(2\pit)')	Display a legend on the figure

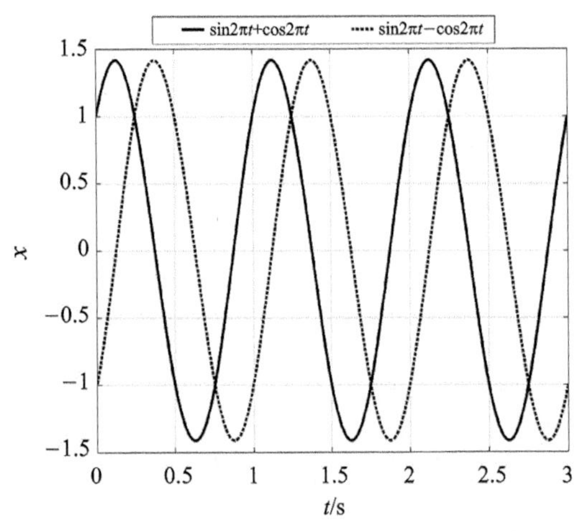

Fig. 2. 5　Graph for simple harmonic motions

It is easy to transfer Eq. (2.10) into Eq. (2.14) in MATLAB by using the function atan2. For both functions $x(t) = \sin 2\pi t + \cos 2\pi t$ and $y(t) = \sin 2\pi t - \cos 2\pi t$, it is easy to find that both curves have amplitude $R = \sqrt{2}$, the phase angles δ can be calculated by MATLAB function atan2. To plot these functions in form of Eq. (2.14), the following MATLAB codes can be used to compare the results between Eq. (2.10) and Eq. (2.14), the results is shown in Fig. 2.6. From Fig. 2.6, it confirms that Eq. (2.10) and Eq. (2.14) can obtain the same results.

Line	MATLAB Codes	Comments
1	clear all, close all	Remove all variables in workspace and close all figures
2	t = linspace(0, 3, 1e3);	To generates linearly spaced vectors
3	x = sin(2 * pi * t) + cos(2 * pi * t);	Calculate $x(t) = \sin 2\pi t + \cos 2\pi t$
4	y = sin(2 * pi * t)−cos(2 * pi * t);	Calculate $y(t) = \sin 2\pi t - \cos 2\pi t$
5	R=sqrt(2);	Calculate the amplitude for x and y
6	ang1=atan2(1,1);	P = atan2(Y,X) returns an array P the same size as X and Y containing the element-by-element, four-quadrant inverse tangent (arctangent) of Y and X, which must be real
7	x1=R * sin(2 * pi * t+ang1);	Calculate x by using Eq. (2.13)
8	ang2=atan2(1,−1);	Calculate the phase angle
9	y1=R * sin(2 * pi * t+ang2);	Calculate y by using Eq. (2.13)
10	figure(2)	Create a new figure
11	plot(t, x,'k', 'linewidth', 2)	Plot the x by solid line (black color and linewidth=2 points)
12	hold on	Retain the current plot for multiple-line plot
13	plot(t, x1,'b:', 'linewidth', 2)	Plot the x1 by dashed line (blue color and linewidth=2 points)
14	grid on	To displays the major grid lines for the current axes
15	xlabel ('{\itt} (Second)', 'fontsize', 14, 'fontname', 'times new roman')	Label the x-axis
16	ylabel ('{\itx}', 'fontsize', 14, 'fontname','times new roman')	Label the y-axis
17	legend('sin(2\pit) + cos(2\pit)', 'sqrt(2) * sin(2\pit+\pi/4)')	Display a legend on the figure
18	figure(3)	Create a new figure
19	plot(t, y,'k', 'linewidth', 2)	Plot the y by solid line (black color and linewidth=2 points)
20	hold on	Retain the current plot for multiple-line plot
21	plot(t, y1,'b:', 'linewidth', 2)	Plot the y1 by dashed line (blue color and linewidth=2 points)

(continued)

22	grid on	To displays the major grid lines for the current axes
23	xlabel (' { \ itt } (Second) ', 'fontsize', 14, 'fontname', 'times new roman')	Label the x -axis
24	ylabel (' { \ itx } ', 'fontsize', 14, 'fontname','times new roman')	Label the y -axis
25	legend(' — sin (2\ pit) + cos (2\ pit)','sqrt(2) * sin(2\pit-\pi/4)')	Display a legend on the figure

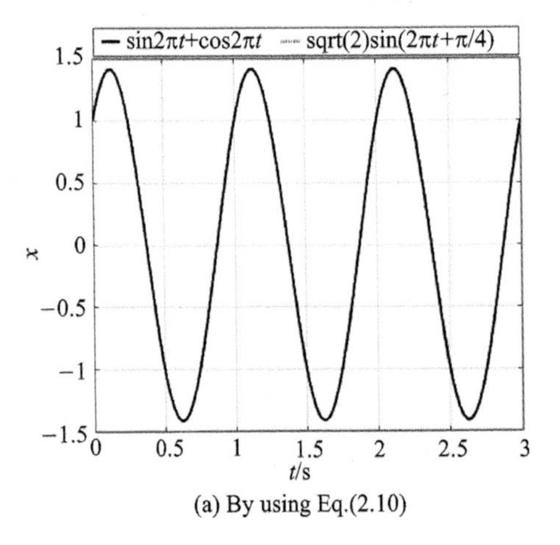

(a) By using Eq.(2.10)

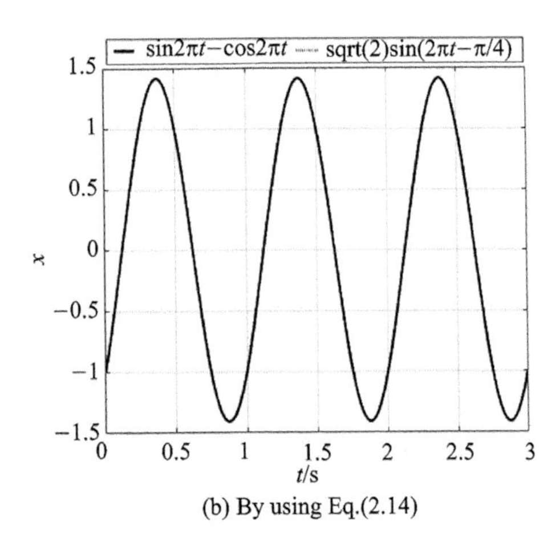

(b) By using Eq.(2.14)

Fig. 2.6 The graph for simple harmonic motions by using Eqs. (2.10) and (2.14)

2.2.4 Solution for undamped SDOF systems

The coefficients A and B in Eq. (2.10) can be determined by using the initial conditions. In general, we assume that

$$x(t=0)=x_0 \tag{2.15a}$$

and

$$\dot{x}(t=0)=v_0 \tag{2.15b}$$

are the initial displacement and initial velocity of the mass m, respectively.

Substituting Eq. (2.15a) into Eq. (2.10), we have

$$B=x_0 \tag{2.15c}$$

Differentiating Eq. (2.10) with respect to time t, we get

$$\dot{x}(t)=\omega_n\left[A\cos(\omega_n t)-B\sin(\omega_n t)\right] \tag{2.16}$$

Substituting Eq. (2.15b) into Eq. (2.16), we have

$$v_0=A\omega_n \quad \text{and} \quad A=\frac{v_0}{\omega_n} \tag{2.17}$$

The solution for Eq. (2.9) is finally determined by combining Eqs. (2.10), (2.15) and (2.17), yields

$$x(t) = x_0 \cos(\omega_n t) + \frac{v_0}{\omega_n} \sin(\omega_n t) \tag{2.18}$$

with $\omega_n = \sqrt{\frac{k}{m}}$.

The mass-spring system is a basic model that represents many other mathematically equivalent SDOF systems. Consider the following example.

Example 2.1

Assume that a mass m is attached at the end of a cantilever beam, and assume that the beam has negligible mass, as shown in Fig. 2.7. Ignore gravity, the mass is assumed as a particle and vibrate with small amplitude about the static equilibrium position. It is well-known that the maximal deflection of the beam under a concentrated force F applied at free end is

$$\Delta = \frac{FL^3}{3EI} \tag{2.19}$$

where E is the Young's modulus and I is the moment of inertia of the cross-section of the beam.

Determine the equation of motion for this vibration system.

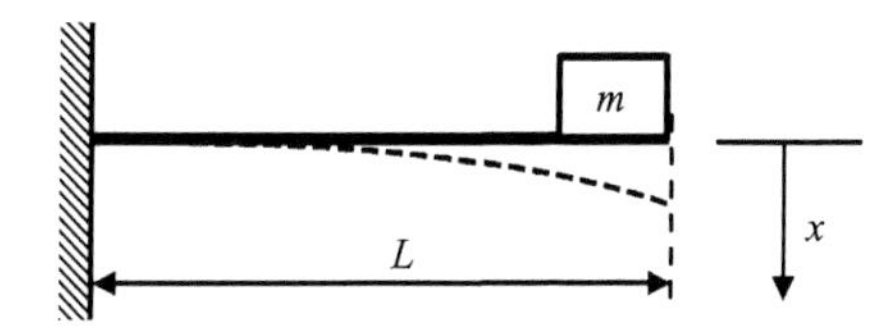

Fig. 2.7 A mass m attached to the end of a cantilever beam (assume that the beam has negligible mass)

Solution:

Notice that Δin Eq. (2.19) is the static deflection of the beam, Eq. (2.19), and the beam can be seen as a spring with constant stiffness k, so the value of stiffness in this case can be expressed as

$$k = \frac{F}{\Delta} = \frac{3EI}{L^3} \tag{2.20}$$

Substituting Eq. (2.20) into Eq. (2.9), it is easy to find that the differential equations of motion for the mass m, measured from the static equilibrium position can be expressed as

$$m\ddot{x} + \frac{3EI}{L^3} x = 0 \tag{2.21}$$

2.3 Rotary Systems

The equation of motion (Eq. (2.9)) was derived from the Newton's second law. We may similarly analyze rotary (or torsional) vibration systems shown in Fig. 2.8 by using the moment-rotation relationship.

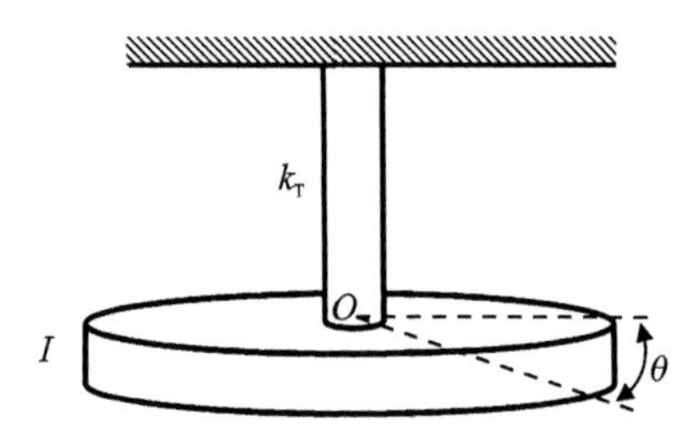

Fig. 2.8 The rotary SDOF system

For a rigid body in general motion, we have

$$\sum M_G = I_G \ddot{\theta} \tag{2.22}$$

where M_G and I_G are the moment and moment of inertia of the body about the center of gravity G, respectively, and θ is the angle displacement. Notice that the disk rotates about a fixed point O in Fig. 2.8, we may rewrite Eq. (2.22) as

$$M_O = I_O \ddot{\theta} \tag{2.23}$$

where M_O and I_O are the moment and moment of inertia of the body about the fixed point O, respectively.

On the other hand, when the disk with moment of inertia I_O is rotated through an angle θ from its equilibrium position, a moment M_O develops between the disk and the shaft. Thus, the shaft acts as a torsional springe of stiffness k_T, that is

$$M_O = -k_T \theta \tag{2.24}$$

where k_T is a positive constant. By using Eqs. (2.23) and (2.24), the following equation of motion can be obtained:

$$I\ddot{\theta} + k_T \theta = 0 \tag{2.25}$$

with the initial conditions

$$\theta(0) = \theta_0 \quad \text{and} \quad \dot{\theta}(0) = \dot{\theta}_0 \tag{2.26}$$

Clearly, we have in analogy to Eq. (2.18), such as

$$\theta(t) = \theta_0 \cos(\omega_n t) + \frac{\dot{\theta}_0}{\omega_n} \sin(\omega_n t) \tag{2.27}$$

with $\omega_n = \sqrt{\dfrac{k_T}{I}}$.

Example 2.2

Assume that a rigid beam with mass M is horizontal at static equilibrium position, as shown in Fig. 2.9. The distance between the support point O and the Center of Gravity (CG) is l, and $k_1 = k_2 = k$. The moment of inertia of the beam about the fixed point O is assumed as I_O. For small vibrations about the equilibrium position (rotating around support point O), determine the equation of motion for the beam.

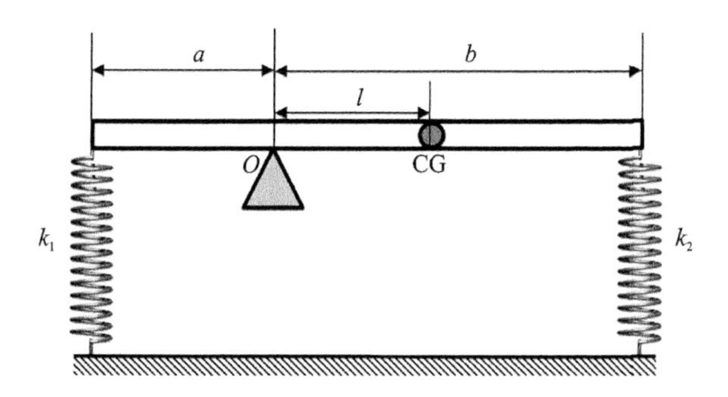

Fig. 2.9 A horizontal beam at static equilibrium position

Solution:

First, the free body diagram is shown in Fig. 2.10. Assume that F_1 and F_2 are the spring forces at static equilibrium. Notice that the beam is horizontal at this position, taking moments about O gives

$$F_2 b = Mgl + F_1 a \tag{2.28}$$

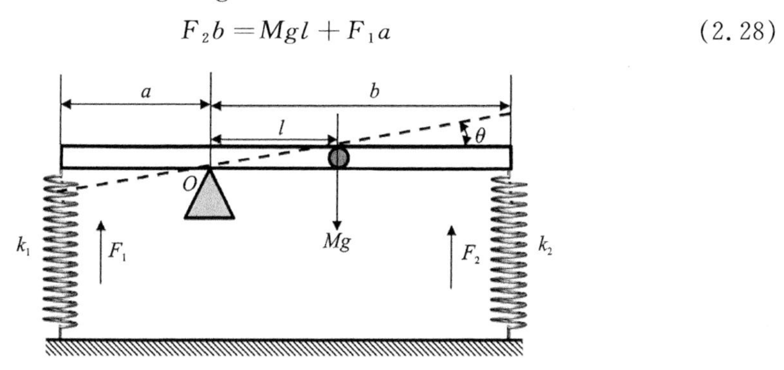

Fig. 2.10 Free body diagram for Example 2.2

If the beam rotates a small angle θ, the springs k_1 and k_2 will be deformed $ka\sin\theta$ and $kb\sin\theta$. According to small angle assumption, we have $\sin\theta = \theta$, the moment about the fixed point O can be expressed as

$$\sum M_O = (F_2 - kb\theta)b - Mgl - (F_1 + ka\theta)a = I_O\ddot{\theta} \tag{2.29}$$

By using Eq. (2.28), Eq. (2.29) can be rewritten as

$$-kb^2\theta - ka^2\theta = I_O\ddot{\theta} \tag{2.30a}$$

or

$$I_O\ddot{\theta} + k(b^2 + a^2)\theta = 0 \tag{2.30b}$$

Eq. (2.30b) is the motion equation for Example 2.2.

2.4 Springs Connected in Series or in Parallel

Sometimes more than one spring acts in a vibration system. The spring, which is considered to be an elastic element of constant stiffness, can take many forms in practice. For example, it can be a wire coil, rubber block, beam or air bag. Springs can be combined in series, parallel and in a combination of series and parallel. It should be noticed that the combined spring units can be replaced in the analysis by a single spring of equivalent stiffness.

When springs are in series, they experience the same force but undergo different deflections. For the two systems to be equivalent, the total static deflection of the original and the equivalent system must be the same. A two-spring system shown in Fig. 2.11(a) can be replaced by the equivalent spring k_{eq} in Fig. 2.11(b).

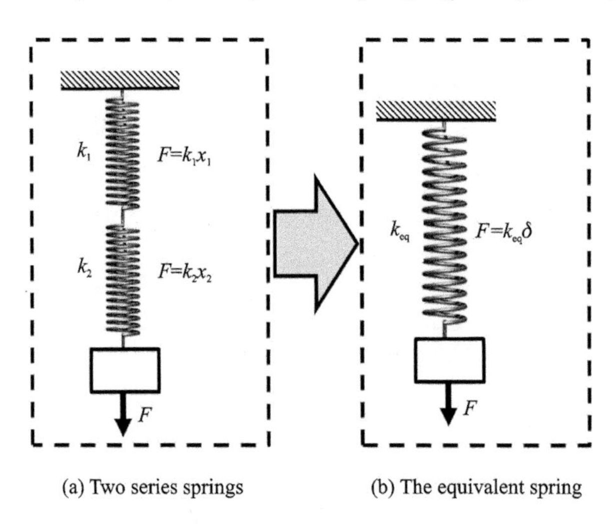

Fig. 2.11 Springs in series

If the deflection at the free end, δ, experienced by applying the force F is to be the same in both cases, so

$$\delta = \frac{F}{k_1} + \frac{F}{k_2} = \frac{F}{k_{eq}} \tag{2.31}$$

From Eq. (2.31), it can be found that

$$\frac{1}{k_{eq}} = \frac{1}{k_1} + \frac{1}{k_2} \quad \text{and} \quad k_{eq} = \frac{k_1 k_2}{k_1 + k_2} \tag{2.32a}$$

In general, the reciprocal of the equivalent stiffness of springs connected in series is obtained by summing the reciprocal of the stiffness of each spring. If there are N springs connected in series, we have

$$\frac{1}{k_{eq}}=\sum_{n=1}^{N}\frac{1}{k_n} \tag{2.32b}$$

Two springs connected in parallel of Fig. 2.12(a) can be replaced by the equivalent spring of Fig. 2.12(b).

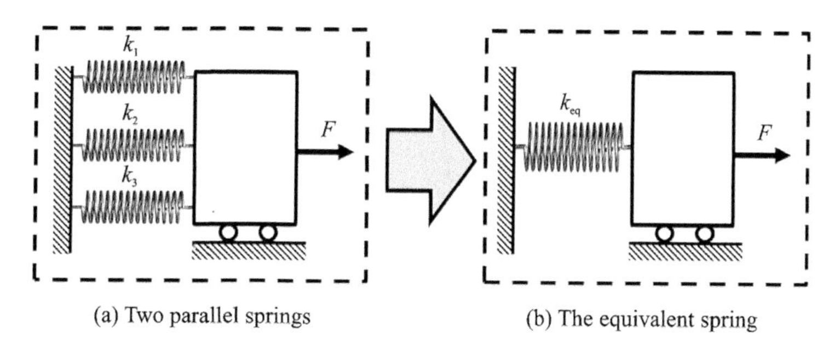

(a) Two parallel springs (b) The equivalent spring

Fig. 2.12 Two springs connected in parallel

It can be found that the defection δ must be the same in both cases, and the sum of the forces exerted by the springs in parallel must equal the force exerted by the equivalent spring. Thus

$$F=k_1\delta+k_2\delta=k_{eq}\delta \tag{2.33}$$

From Eq. (2.33), it is easy to find that

$$k_{eq}=k_1+k_2 \tag{2.34a}$$

In general, the equivalent stiffness of springs connected in parallel is obtained by summing the stiffness of each spring. If there are N springs connected in series, we have

$$k_{eq}=\sum_{n=1}^{N}k_n \tag{2.34b}$$

2.5 Modelling Using Energy Method

According to Eqs. (2.10) and (2.13) and Subsection 2.2.2, it is easy to find that the general solution of

$$m\ddot{x}+kx=0 \tag{2.35}$$

can be expressed alternatively in the form

$$x(t)=C\sin(\omega_n t+\delta) \tag{2.36}$$

where C and δ are two arbitrary constants and $\omega_n=\sqrt{\frac{k}{m}}$.

Let t_1 and t_2 are the times associated with the maximum and zero values of the solution $x(t)$, respectively, as shown in Fig. 2.13. Denote $x_1 = x(t_1)$ and $x_2 = x(t_2)$. By using Eq. (2.36), it can be found that

$$x_1 = C \quad \text{and} \quad \dot{x}_1 = 0 \tag{2.37a}$$

$$x_2 = 0 \quad \text{and} \quad \dot{x}_2 = C\omega_n \tag{2.37b}$$

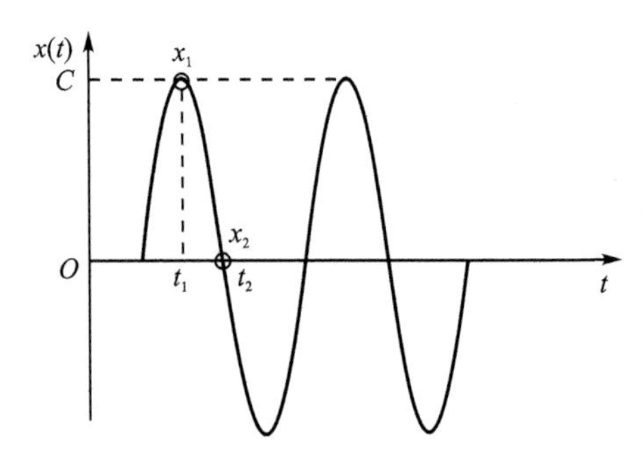

Fig. 2.13 The response of the system

From Eqs. (2.37a) and (2.37b), it is easy to find that

$$\frac{\dot{x}_2^2}{x_1^2} = \omega_n^2 \tag{2.38}$$

Assume that U and T are the potential energy and kinetic energy of the vibration system, respectively. The potential energy U is always reference at the static equilibrium position $x = 0$. From Fig. 2.13, it is easy to find that at t_1 position the kinetic energy is zero and the potential energy is maximum:

$$T_1 = 0 \quad \text{and} \quad U_1 = U_{\max} \tag{2.39a}$$

Similarly, at t_2 position, we get

$$T_2 = T_{\max} \quad \text{and} \quad U_2 = 0 \tag{2.39b}$$

Notice that the mechanical energy is conserved for any undamped system. Conservation of energy implies that

$$T + U = \text{constant} \quad \text{and} \quad T_{\max} = U_{\max} \tag{2.40}$$

The potential energy T is a function of velocity and the potential energy is usually a function of displacement. Hence ω_n can be obtained by using Eq. (2.38) or Eq. (2.40).

Procedure for energy method modelling

(1) Determine the maximal potential energy $U_1 = U_1(x_1^2)$ (where $T_1 = 0$) and the maximal kinetic energy $T_2 = T_2(\dot{x}_2^2)$ (where $U_2 = 0$).

(2) By using Eq. (2.38) or Eq. (2.40) determine the natural frequency, such as $\omega_n^2 = \frac{\dot{x}_2^2}{x_1^2}$, $T_{\max} = U_{\max}$ or $T + U = \text{constant}$.

Example 2.3

Determine the equation of motion for the system shown in Fig. 2.14. Assume that the disks A and B are fixed together with mass m, and the disks can rotate at point O without friction. The moment of inertia of the disk about the point O is assumed as I_O.

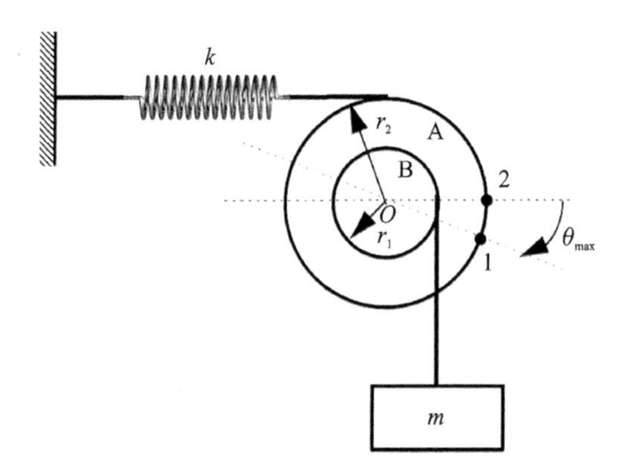

Fig. 2.14 SDOF system for Example 2.3

Solution:

Assume that the static equilibrium position is at point 2, and at point 1 is the zero angle velocity point (or maximal angle displacement point). So the point 1 has the maximal potential energy location.

$$U_{max} = U_1 = \frac{1}{2}k(r_2\theta_1)^2 = \frac{1}{2}kr_2^2\theta_1^2 \tag{2.41a}$$

At point 2 (the static equilibrium point with maximal kinetic energy)

$$T_{max} = T_2 = \frac{1}{2}I_O\dot{\theta}_2^2 + \frac{1}{2}m(r_1\dot{\theta}_2)^2 = \frac{1}{2}(I_O + mr_1^2)\dot{\theta}_2^2 \tag{2.41b}$$

By using Eq. (2.40) $U_{max} = T_{max}$, we get

$$\frac{1}{2}kr_2^2\theta_1^2 = \frac{1}{2}(I_O + mr_1^2)\dot{\theta}_2^2 \tag{2.42}$$

From Eq. (2.42), it can be found that

$$\frac{\dot{\theta}_2^2}{\theta_1^2} = \frac{kr_2^2}{I_O + mr_1^2} \tag{2.43}$$

By using Eq. (2.38), we get

$$\omega_n^2 = \frac{kr_2^2}{I_O + mr_1^2} \tag{2.44}$$

According to the general formula for equation of motion of the SDOF system, the motion equation can be written as

$$\ddot{\theta} + \omega_n^2\theta = \ddot{\theta} + \frac{kr_2^2}{I_O + mr_1^2}\theta = 0 \tag{2.45a}$$

Eq. (2.45a) can also be rewritten as

$$(I_O + mr_1^2)\ddot{\theta} + kr_2^2\theta = 0 \tag{2.45b}$$

Remark 1 The analysis above ignores gravity effects. In the presence of gravity the system still has the same natural frequency, as shown in Eq. (2.44). Also, the motion Eq. (2.45) holds for this case. Accounting gravity is only changing the static equilibrium position.

Remark 2 There is another way to use energy method. Notice that there is no energy dissipation in undamped systems, which contain energy storage elements only. In other words, energy is conserved in these systems, which are known as conservative systems. For mechanical systems, conservation of energy gives $T + U =$ constant (i. e., Eq. (2.40)), it is easy to find that

$$\frac{\mathrm{d}(T+U)}{\mathrm{d}t} = 0 \tag{2.46}$$

The potential energy T and kinetic energy U at any angle θ can be written as

$$U = \frac{1}{2}k(r_2\theta)^2 \quad \text{and} \quad T = \frac{1}{2}I_O\dot{\theta}^2 + \frac{1}{2}m(r_1\dot{\theta})^2 \tag{2.47}$$

Substituting Eq. (2.47) into Eq. (2.46), we get

$$\frac{\mathrm{d}(T+U)}{\mathrm{d}t} = \frac{\mathrm{d}\left[\frac{1}{2}k(r_2\theta)^2 + \frac{1}{2}I_O\dot{\theta}^2 + \frac{1}{2}m(r_1\dot{\theta})^2\right]}{\mathrm{d}t}$$

$$= [(I_O + mr_1^2)\ddot{\theta} + kr_2^2\theta]\dot{\theta} = 0 \tag{2.48}$$

In Eq. (2.48), it is noticed that $\dot{\theta}$ will not equal to zero at any angle θ. It means that

$$(I_O + mr_1^2)\ddot{\theta} + kr_2^2\theta = 0 \tag{2.49}$$

We get the same motion equation as shown in Eq. (2.45b).

Example 2.4

A disk B with mass m, rolls on disk A without slipping. The moment of inertia of the disk B about the point O is assumed as I_O. Determine the natural frequency of motion for small oscillations about the equilibrium position.

Solution:

From Fig. 2.15, it can be found that thestatic equilibrium point is at position 2 (the potential energy at position 2 is zero), and the zero angle velocity location is position 1. The potential energy at position 1 can be expressed as

$$U_1 = mg(R-r)(1-\cos\theta) \tag{2.50}$$

Using the power-series expansions

$$\cos z = 1 - \frac{z^2}{2!} + \frac{z^4}{4!} - \frac{z^6}{6!} + \cdots + (-1)^n\frac{z^{2n}}{(2n)!} + \cdots \tag{2.51}$$

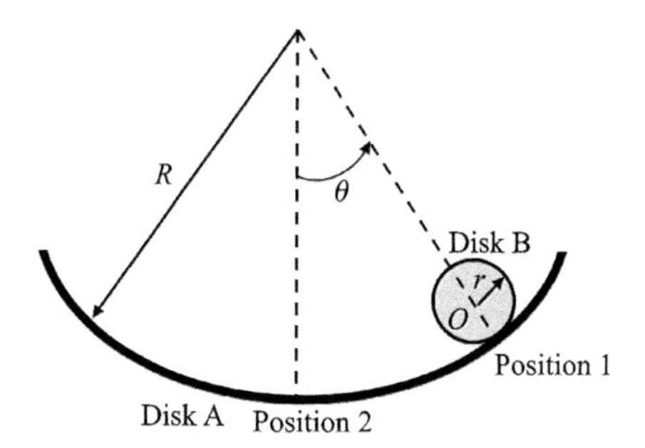

Fig. 2.15 A disk B rolls on disk A without slipping

For a small angle θ_1, we have

$$1-\cos\theta_1 \cong \frac{\theta_1^2}{2!} \tag{2.52}$$

Substituting Eq. (2.52) into Eq. (2.50), the potential energy at position 1 can be rewritten as

$$U_1 = mg(R-r)\frac{\theta_1^2}{2} \tag{2.53}$$

The velocity of the center O of the disk B is $(R-r)\theta$, and the angular velocity of disk B can be expressed as $\frac{R-r}{r}\dot{\theta}$. So the kinetic energy at position 2 (the static equilibrium point) can be expressed as

$$\begin{aligned} T_1 &= \frac{1}{2}m\left[(R-r)\dot{\theta}_2\right]^2 + \frac{1}{2}I_O\left[\frac{(R-r)\dot{\theta}_2}{r}\right]^2 \\ &= \frac{1}{2}(R-r)^2\left(m+\frac{I_O}{r^2}\right)\dot{\theta}_2^2 \end{aligned} \tag{2.54}$$

By using relationship $U_{\max} = T_{\max}$, it is easy to obtain

$$mg(R-r)\frac{\theta_1^2}{2} = \frac{1}{2}(R-r)^2\left(m+\frac{I_O}{r^2}\right)\dot{\theta}_2^2 \tag{2.55}$$

From Eq. (2.55), the natural frequency can be obtained by using Eq. (2.38), that is

$$\omega_n^2 = \frac{\dot{\theta}_2^2}{\theta_1^2} = \frac{mg}{(R-r)\left(m+\frac{I_O}{r^2}\right)} \tag{2.56}$$

Example 2.5

Assume that the mass of the spring shown in Fig. 2.16 is m_s, and the spring mass is evenly distributed along the length. Approximate the natural frequency of the system taking into account the mass of the spring. Assume small oscillations about the static equilibrium position.

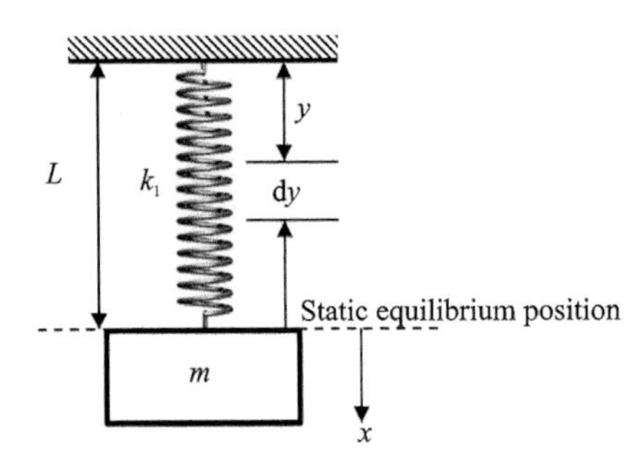

Fig. 2.16 The system taking into account the mass of the spring

Solution:

Notice that gravity has no effect on the natural frequency for mass-spring vibration system, if the coordinate origin is the static equilibrium position, as shown in Fig. 2.16. Hence, ignoring gravity, the potential energy at position x is

$$U=\frac{1}{2}kx^2 \tag{2.57}$$

The kinetic energy at x consists of two components. The kinetic energy depends on the motion of the mass m and the kinetic energy of the spring k. The kinetic energy of the mass can be simply written as

$$T^m=\frac{1}{2}m\dot{x}^2 \tag{2.58}$$

We assume that the distribution of the velocity along the spring is by a linear function, such as

$$\dot{y}=\frac{y}{L}\dot{x} \tag{2.59}$$

By using Eq. (2.59), it can be found that the kinetic energy for the spring element dy is

$$T^{dy}=\frac{1}{2}\frac{m_s dy}{L}\left(\frac{y}{L}\dot{x}\right)^2 \tag{2.60}$$

The kinetic energy of the total spring is obtained by integrating:

$$T^s=\int_0^L T^{dy}dy=\int_0^L \frac{1}{2}\frac{m_s}{L}\left(\frac{\dot{x}}{L}y\right)^2 dy=\frac{m_s\dot{x}^2y^3}{6L^3}\Bigg|_{y=0}^{L}=\frac{m_s\dot{x}^2}{6} \tag{2.61}$$

From Eqs. (2.58) and (2.61), the total kinetic energy can be written as

$$T=T^m+T^s=\frac{1}{2}\left(m+\frac{m_s}{3}\right)\dot{x}^2 \tag{2.62}$$

By using relationship $U+T=$constant, it is easy to obtain

$$\frac{d(U+T)}{dt}=\frac{d\left[\frac{1}{2}kx^2+\frac{1}{2}\left(m+\frac{m_s}{3}\right)\dot{x}^2\right]}{dt}=\left(m+\frac{m_s}{3}\right)\ddot{x}+kx=0 \tag{2.63}$$

From Eq. (2.63), the natural frequency can be obtained as

$$\omega_{\mathrm{n}}^{\text{with spring mass}} = \sqrt{\frac{k}{m+\frac{m_{\mathrm{s}}}{3}}} = \sqrt{\frac{k}{m}}\sqrt{\frac{1}{1+\frac{m_{\mathrm{s}}}{3m}}} \tag{2.64}$$

From Eq. (2.64), it can be found that the natural frequency will be greater estimated without considering the spring mass, as shown in Fig. 2.17.

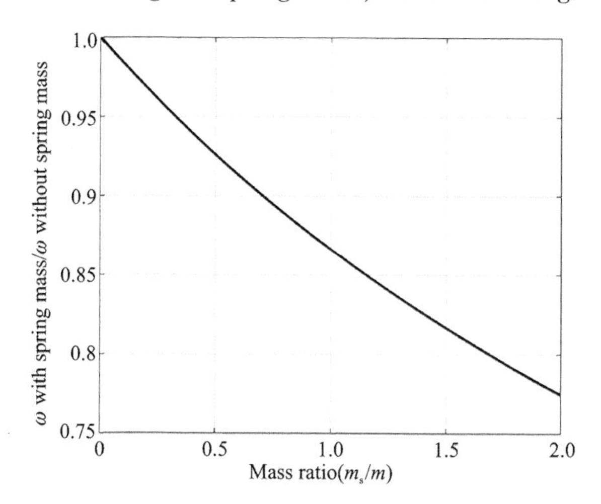

Fig. 2.17 The natural frequency ratio with and without considering the spring mass

2.6 Viscously Damped SDOF Systems

Now consider the free (natural) response of a simple oscillator in the presence of energy dissipation (damping). The free vibration of the mass-spring-damper system is shown in Fig. 2.18. Apply Newton's second law, and from the free body diagram in Fig. 2.18, measured from the static equilibrium position, one has the equation of motion governed by the second order differential equation:

$$m\ddot{x} + c\dot{x} + kx = 0 \tag{2.65}$$

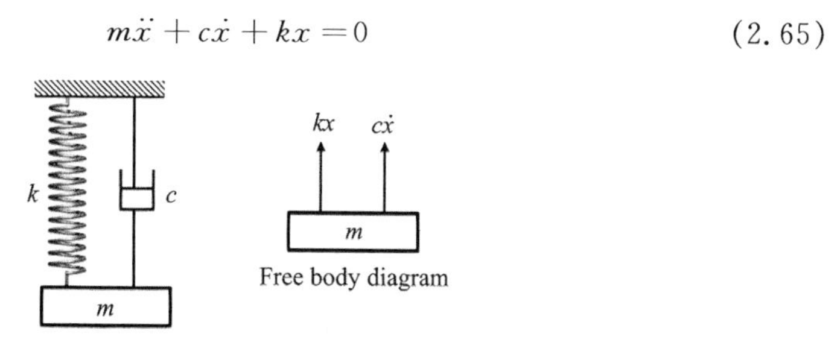

Fig. 2.18 The damped SDOF system

Eq. (2.65) holds also in the present of gravitation field. The viscous damping constant is denoted by c.

The following notations are used in this textbook.

ω_n = undamped natural frequency.

ω_d = damped natural frequency.

ω = frequency of excitation.

Denote natural frequency of undamped system

$$\omega_n^2 = \sqrt{\frac{k}{m}} \tag{2.66}$$

and damping ratio

$$\zeta = \frac{c}{2\sqrt{km}} \tag{2.67}$$

By using Eqs. (2.66) and (2.67), the equation of motion in Eq. (2.65) can be rewritten as

$$\ddot{x} + 2\zeta\omega_n\dot{x} + \omega_n^2 x = 0 \tag{2.68}$$

Assume that Eqs. (2.65) and (2.68) have an exponential solution, such as

$$x = C\exp(\lambda t) \tag{2.69}$$

where C and λ are unknown parameters.

This is justified because the linear systems have exponential or oscillatory (i.e., complex exponential) free responses. A more detailed proof can be found in Appendix A.

Substituting Eq. (2.69) into Eq. (2.68), we get

$$(\lambda^2 + 2\zeta\omega_n\lambda + \omega_n^2)C\exp(\lambda t) = 0 \tag{2.70}$$

Notice that $C\exp(\lambda t)$ in Eq. (2.70) is not zero in general. It means that Eq. (2.71) will represent a solution of Eq. (2.68) only if λ satisfies the equation

$$\lambda^2 + 2\zeta\omega_n\lambda + \omega_n^2 = 0 \tag{2.71}$$

Eq. (2.71) is called the characteristic equation of the system. This equation depends on the natural dynamics of the system only. There is no forcing excitation or initial conditions in Eq. (2.71).

Clearly, there are two roots in the solution of Eq. (2.71), such as

$$\lambda_1 = -\zeta\omega_n - \sqrt{\zeta^2 - 1}\cdot\omega_n \quad \text{and} \quad \lambda_2 = -\zeta\omega_n + \sqrt{\zeta^2 - 1}\cdot\omega_n \tag{2.72}$$

These are called eigenvalues or poles of the system.

When $\lambda_1 \neq \lambda_2$, the general solution for Eq. (2.68) can be written as

$$x = C_1\exp(\lambda_1 t) + C_2\exp(\lambda_2 t) \tag{2.73}$$

The two unknown constants C_1 and C_2 in Eq. (2.73) are related to the integration constants, and can be determined by two initial conditions (such as the displacement and velocity at initial time $t = 0$).

If $\lambda_1 = \lambda_2 = \lambda$, one has the case of repeated roots. In this case, the general solution of Eq. (2.73) does not hold because C_1 and C_2 would no longer be independent constants determined by initial conditions. The repetition of the roots suggests that one term of the homogenous solution should have the multiplier t (a result of the double-integration of zero). Then, the general solution for Eq. (2.71) for $\lambda_1 = \lambda_2 = \lambda$ is

$$x = C_1 \exp(\lambda t) + C_2 t \exp(\lambda t) \tag{2.74}$$

Denote the damped natural frequency

$$\omega_d = \sqrt{1-\zeta^2}\,\omega_n \tag{2.75}$$

Its solution(s) will be either negative real numbers, or complex numbers with negative real parts. The displacement $x(t)$ behaves differently depending on the value of ζ relative to m and k. There are three possible classes of behaviors based on the possible types of root(s) of the characteristic polynomial. The nature of the response will depend on the particular category of damping, as discussed below.

2.6.1 Case 1: Overdamped motion ($\zeta > 1$)

If the damping ratio $\zeta > 1$, in this case, roots λ_1 and λ_2 of the characteristic Eq. (2.71) are real, such as

$$\lambda_1 = -\zeta\omega_n + \omega_n\sqrt{\zeta^2-1} < 0 \tag{2.76a}$$

$$\lambda_2 = -\zeta\omega_n - \omega_n\sqrt{\zeta^2-1} < 0 \tag{2.76b}$$

From Eq. (2.76), it can be clearly found that both λ_1 and λ_2 are negative for this case. Hence, from Eq. (2.73), it is clearly that $x \to 0$ as $t \to \infty$. This means that the system is asymptotically stable.

Recall the initial conditions

$$x(0) = x_0 \quad \text{and} \quad \dot{x}(0) = v_0 \tag{2.77}$$

Substituting Eq. (2.77) into general solution in Eq. (2.73), we get

$$x_0 = C_1 + C_2 \tag{2.78}$$

and

$$v_0 = \lambda_1 C_1 + \lambda_2 C_2 \tag{2.79}$$

Multiply the first initial condition in Eq. (2.78) by λ_1, we get

$$\lambda_1 x_0 = \lambda_1 C_1 + \lambda_1 C_2 \tag{2.80}$$

Then subtract Eq. (2.80) from Eq. (2.79), we obtain

$$v_0 - \lambda_1 x_0 = C_2(\lambda_2 - \lambda_1) \tag{2.81}$$

and

$$C_2 = \frac{v_0 - \lambda_1 x_0}{\lambda_2 - \lambda_1} \tag{2.82}$$

Similarly, multiply the first initial condition in Eq. (2.78) by λ_2 and subtract from Eq. (2.79), one obtains

$$v_0 - \lambda_2 x_0 = C_1(\lambda_1 - \lambda_2) \tag{2.83}$$

hence,

$$C_1 = \frac{v_0 - \lambda_2 x_0}{\lambda_1 - \lambda_2} \tag{2.84}$$

Based on above analysis, substituting Eqs. (2.76), (2.82) and (2.84) into Eq. (2.73), the response of the system can be expressed as

$$x = C_1 \exp\left[(-\zeta + \sqrt{\zeta^2 - 1})\omega_n t\right] + C_2 \exp\left[(-\zeta - \sqrt{\zeta^2 - 1})\omega_n t\right] \tag{2.85}$$

with

$$C_1 = \frac{v_0 + (\zeta + \sqrt{\zeta^2 - 1})\omega_n x_0}{2\omega_n \sqrt{\zeta^2 - 1}} \tag{2.86a}$$

$$C_2 = \frac{-v_0 - (\zeta - \sqrt{\zeta^2 - 1})\omega_n x_0}{2\omega_n \sqrt{\zeta^2 - 1}} \tag{2.86b}$$

A vibration system with displacement function in Eq. (2.85) is called overdamped system. It should be noticed that the system does not oscillate; it has no periodic components in the solution. In fact, depending on the initial conditions the mass of an overdamped vibration system might or might not cross over its equilibrium position. **But it could cross the equilibrium position at most once.**

For example, assume that $C_1 = C_2 = \omega_n = 1$ and damping ratio $\zeta = 2$. Eq. (2.85) can be rewritten as $x = \exp\left[(-2+\sqrt{3})t\right] + \exp\left[(-2-\sqrt{3})t\right]$. It can be calculated by the following MATLAB codes, and we can obtain the results shown in Fig. 2.19.

Line	MATLAB Codes	Comments
1	t=linspace(0,20,1e3);	To generates linearly spaced vectors
2	x=exp((-2+sqrt(3)) * t)+ exp((-2-sqrt(3)) * t);	To calculate the function $x = \exp\left[(-2+\sqrt{3})t\right] + \exp\left[(-2-\sqrt{3})t\right]$
3	plot(t,x,'k')	To plot x
4	xlabel('\itt'), ylabel('\itx')	To add label for x-axis and y-axis

Assume that $C_1 = -1$, $C_2 = 2$, $\omega_n = 1$ and $\zeta = 2$. Eq. (2.85) can be rewritten as $x = -\exp\left[(-2+\sqrt{3})t\right] + 2\exp\left[(-2-\sqrt{3})t\right]$. It can be found that the response will cross over its equilibrium position once, as shown in Fig. 2.20.

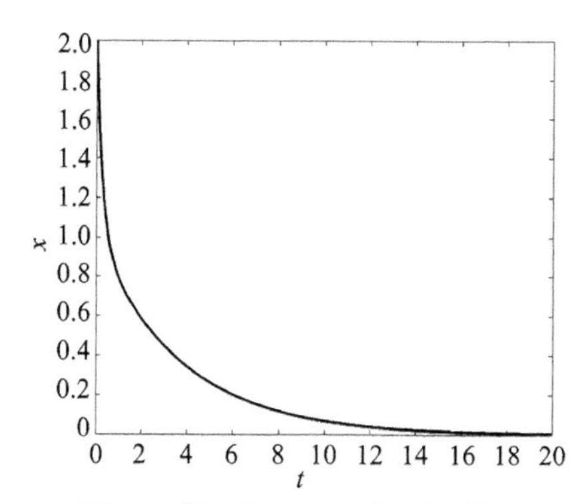

Fig. 2.19 Response for the first overdamped vibration system

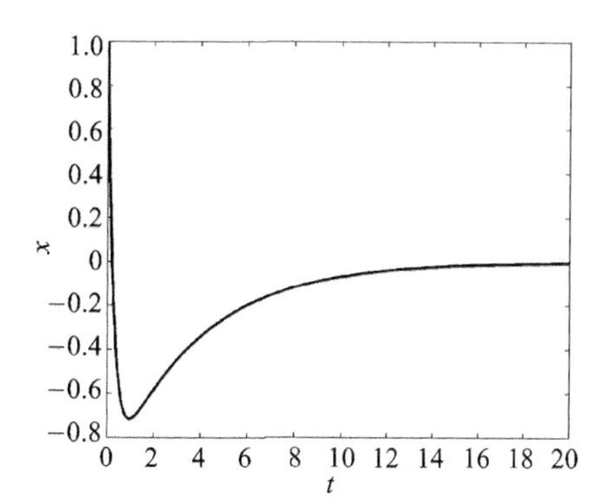

Fig. 2.20 Response for the second overdamped vibration system

2.6.2 Case 2: Underdamped motion ($\zeta < 1$)

In this case, the roots of the characteristic equation in Eq. (2.72) are

$$\lambda_1 = -\zeta\omega_n + j\omega_n\sqrt{\zeta^2 - 1} \quad \text{and} \quad \lambda_2 = -\zeta\omega_n - j\omega_n\sqrt{\zeta^2 - 1} \tag{2.87}$$

where $j = \sqrt{-1}$.

By using Eq. (2.75), Eq. (2.87) can be rewritten as

$$\lambda_1 = -\zeta\omega_n + j\omega_d \quad \text{and} \quad \lambda_2 = -\zeta\omega_n - j\omega_d \tag{2.88}$$

Notice that λ_1 and λ_2 are complex conjugates in Eq. (2.88). Substituting Eq. (2.88) into Eq. (2.73), the response for Eq. (2.53) in this case can be expressed as

$$x = \exp(-\zeta\omega_n t)\left[C_1\exp(j\omega_d t) + C_2\exp(-j\omega_d t)\right] \tag{2.89}$$

The term within the square brackets of Eq. (2.89) has to be real because it represents the time response of a real physical system. It means that C_1 and C_2 should also be complex conjugates.

Notice that

$$\exp(j\omega_d t) = \cos(\omega_d t) + j\sin(\omega_d t) \tag{2.90a}$$

$$\exp(-j\omega_d t) = \cos(\omega_d t) - j\sin(\omega_d t) \tag{2.90b}$$

Thus an alternative form of the general solution in Eq. (2.89) would be

$$x = \exp(-\zeta\omega_n t)\left[A_1\cos(\omega_d t) + A_2\sin(\omega_d t)\right] \tag{2.91}$$

Here, A_1 and A_2 are the two unknown constants. By equating the coefficients, they can be shown that

$$A_1 = C_1 + C_2, \quad A_2 = j(C_1 - C_2) \tag{2.92}$$

hence

$$C_1 = \frac{A_1 - jA_2}{2}, \quad C_2 = \frac{A_1 + jA_2}{2} \tag{2.93}$$

Clearly, C_1 and C_2 are complex conjugates, as required.

Initial Conditions: Let define $x(0)=x_0$, $\dot{x}(0)=v_0$ as before. Substituting these initial conditions into Eq. (2.91), we have

$$x_0 = A_1 \tag{2.94}$$

and

$$v_0 = -\zeta\omega_n A_1 + \omega_d A_2 \tag{2.95}$$

or

$$A_2 = \frac{v_0 + \zeta\omega_n x_0}{\omega_d} \tag{2.96}$$

According to Eq. (2.91), there is another form of the solution:

$$x = A\exp(-\zeta\omega_n t)\sin(\omega_d t + \phi) \tag{2.97}$$

Here, A and ϕ are the unknown constants with

$$A = \sqrt{A_1^2 + A_2^2} \quad \text{and} \quad \sin\phi = \frac{A_1}{\sqrt{A_1^2 + A_2^2}} \tag{2.98}$$

aslo

$$\cos\phi = \frac{A_2}{\sqrt{A_1^2 + A_2^2}} \quad \text{and} \quad \tan\phi = \frac{A_1}{A_2} \tag{2.99}$$

Notice that the response $x \to 0$ as $t \to \infty$. It means that the system is asymptotically stable. A schematic of the free vibration responses of an underdamped SDOF system for various values of $0<\zeta<1$ is shown in Fig. 2.21.

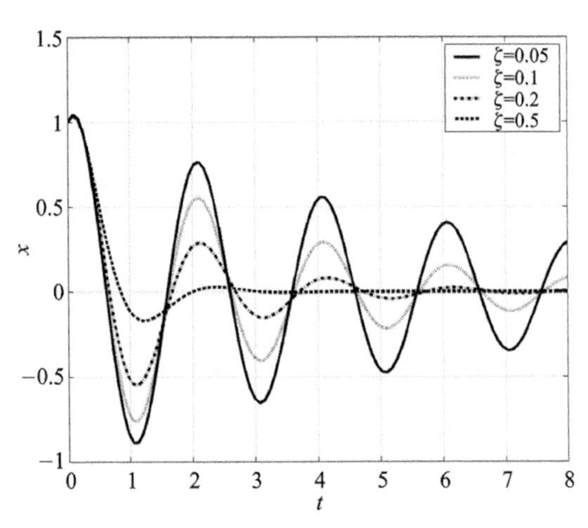

Fig. 2.21 Schematic of the response of an underdamped vibration system

2.6.3 Case 3: Critically damped motion ($\zeta=1$)

For damping ratio $\zeta=1$, recall Eq. (2.72), it can be found that there are two repeated roots, given by

$$\lambda_1 = \lambda_2 = -\omega_n \tag{2.100}$$

The response for this case can be expressed as (similar to Eq. (2.74))

$$x = C_1 \exp(-\omega_n t) + C_2 t \exp(-\omega_n t) \tag{2.101}$$

Since the term $\exp(-\omega_n t)$ goes to zero faster when t goes to infinity, one has

$$t\exp(-\omega_n t) \to 0 \quad \text{as} \quad t \to \infty \tag{2.102}$$

Hence, the system is asymptotically stable.

Now use the initial conditions $x(0)=x_0$, $\dot{x}(0)=v_0$. One obtains

$$x_0 = C_1 \tag{2.103a}$$

$$v_0 = -\omega_n C_1 + C_2 \tag{2.103b}$$

hence,

$$C_2 = v_0 + \omega_n x_0 \tag{2.104}$$

Substituting Eqs. (2.103a) and (2.104) into Eq. (2.101), the response for damping ratio $\zeta=1$ can be expressed as

$$x = x_0 \exp(-\omega_n t) + (v_0 + \omega_n x_0) t \exp(-\omega_n t) \tag{2.105}$$

Note: When $\zeta=1$, one has the critically damped response because below this value, the response is oscillatory (underdamped), and above this value, the response is nonoscillatory (overdamped). It follows that one can define the damping ratio as

$$\zeta = \frac{\text{Damping constant}}{\text{Damping constant for critically damped condition}} \tag{2.106}$$

In Eq. (2.106), the damping constant for critically damped condition is denoted as **critical damping** c_c.

$$c_c = 2\sqrt{mk} \tag{2.107}$$

By using Eq. (2.107), the damping ratio in Eq. (2.67) can be rewritten as

$$\zeta = \frac{c}{c_c} \tag{2.108}$$

From Eq. (2.108), it is now clear why ζ is called as damping ratio. It follows from Eq. (2.108) that if $\zeta < 1$ then the mass-spring-damper oscillates with frequency w_d. We notice that $x \to 0$ as $t \to \infty$, similar to the motion of physical objects. There is a non-oscillatory motion when $\zeta \geqslant 1$.

A schematic of a critically damped response with different initial conditions is shown in Fig. 2.22. A system exhibits this behavior is called critically damped. That is, the damping constant c is just large enough to prevent oscillation. As can be seen, this system does not oscillate, either. Just like the overdamped case, the mass could cross its equilibrium position at most one time.

Comment: The value $c_c = 2\sqrt{mk}$ is called critical damping. It is the threshold level below which damping would be too small to prevent the system from oscillating.

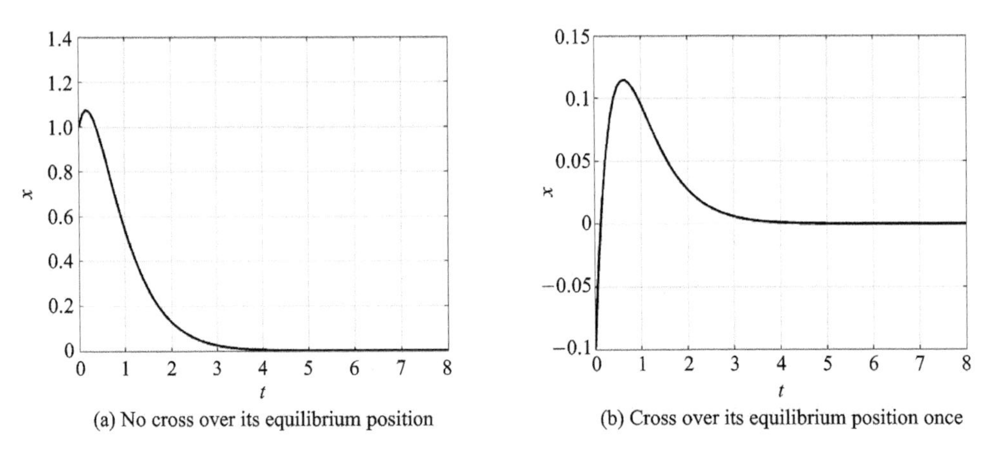

(a) No cross over its equilibrium position (b) Cross over its equilibrium position once

Fig. 2.22 Response of a critically damped vibration system

2.7 Evaluating Damping Ratio from Measurements (Logarithmic Decrement)

Consider the case of damped oscillatory vibration, i. e., $\zeta < 1$. The general solution of Eq. (2.89) for this case can be written alternatively in the form (as shown in Eq. (2.97)):

$$x = A\exp(-\zeta\omega_{\mathrm{n}}t)\sin(\omega_{\mathrm{d}}t + \phi) \tag{2.109}$$

The formulas for A and ϕ are the same as in the Subsection 2.6.2 (Underdamped motion). The displacement function x is oscillating, but the amplitude of oscillation, $A\exp(-\zeta\omega_{\mathrm{n}}t)$, is decaying exponentially, as shown in Fig. 2.21. For all particular solutions (except the zero solution that corresponds to the initial conditions $x(0) = 0$, $\dot{x}(0)=0$), the mass crosses its equilibrium position infinitely often, as expected.

The displacement of an underdamped mass-spring-damper system shows periodic-like motion. However, it is not truly periodic because its amplitude is ever decreasing. It means that it cannot exactly repeat itself. So Eq. (2.109) is a quasi-periodic function.

Notice that the underdamped system is oscillating at damped natural frequency (defined in Eq. (2.75)). The peak-to-peak time of the oscillation is the quasi-period.

$$T_{\mathrm{q}} = \frac{2\pi}{\omega_{\mathrm{d}}} \text{ (seconds)} \tag{2.110}$$

In addition to cause the amplitude to gradually decay to zero, damping has another effect on the oscillating motion, that is, it immediately decreases the quasi-frequency and lengthens the quasi-period (compare to the natural frequency and natural period of an undamped system). The larger the damping constant c (or damping ratio ζ), the smaller quasi-frequency and the longer quasi-period become. Eventually, at the critical damping threshold, when $c_c=2\sqrt{mk}$ and $\zeta=1$, recall Eq. (2.75), we get $\omega_d=\sqrt{1-\zeta^2}\,\omega_n=0$ at the critical damping case. From Eq. (2.110), it can be found that the quasi-frequency tends to infinite and the displacement becomes aperiodic (becoming instead a critically damped system).

Notice that in all three cases of damped free vibration, the displacement tends to zero as time t tends to infinite. This behavior makes perfect sense from a conservation of energy point-of-view: while the system is in motion, the damping wastes away whatever energy the system has started out with, but there is no forcing function to supply the system with additional energy. Consequently, eventually the motion comes to stop.

Denote by t_1 and t_2 as two successive times when the system reach local maximal response. Assume that $x_1=x(t_1)$ and $x_2=x(t_2)$ as shown in Fig. 2.23.

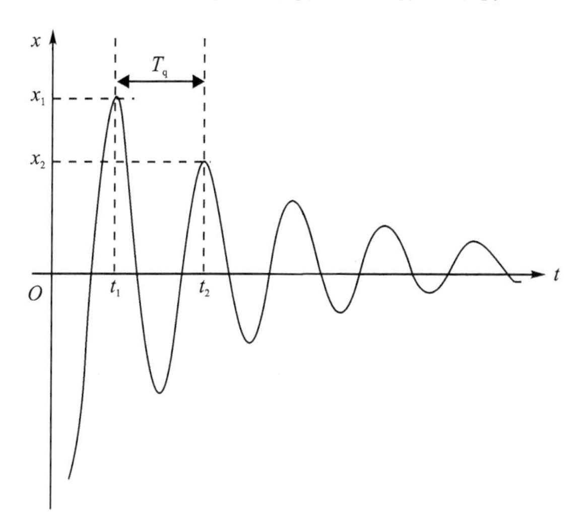

Fig. 2.23 The response of a damped SDOF system

Then due to the choice of t_1 and t_2, according to Eq. (2.109), we have

$$\sin(\omega_d t_1+\phi)=1 \tag{2.111}$$

and

$$\sin(\omega_d t_2+\phi)=1 \tag{2.112}$$

Furthermore, since $t_2=t_1+T_q$ (see Fig. 2.23), by using Eqs. (2.110) and (2.111), it is easy to find that

$$[\omega_d(t_1+T_q)+\phi]-(\omega_d t_1+\phi)=\omega_d T_q=2\pi \tag{2.113}$$

Hence, using Eq. (2.75), such as $\omega_d=\sqrt{1-\zeta^2}\,\omega_n$, Eq. (2.113) can be rewritten as

$$T_q = \frac{2\pi}{\sqrt{1-\zeta^2}\,\omega_n} \tag{2.114}$$

Now substituting Eqs. (2.110) and (2.111) into Eq. (2.109), we get

$$x_1 = A\exp(-\zeta\omega_n t_1) \tag{2.115}$$

$$x_2 = A\exp[-\zeta\omega_n(t_1 + T_q)] \tag{2.116}$$

From Eqs. (2.115) and (2.116), we have

$$\frac{x_1}{x_2} = \frac{A\exp(-\zeta\omega_n t_1)}{A\exp[-\zeta\omega_n(t_1 + T_q)]} = \exp(\zeta\omega_n T_q) \tag{2.117}$$

Define logarithmic decrement

$$\delta = \ln\frac{x_1}{x_2} \tag{2.118}$$

and then substituting Eq. (2.117) into Eq. (2.118), we get

$$\delta = \ln[\exp(\zeta\omega_n T_q)] = \zeta\omega_n T_q \tag{2.119}$$

and by using Eq. (2.114),Eq. (2.119) can be rewritten as

$$\delta = \zeta\omega_n T_q = \frac{2\pi\zeta\omega_n}{\sqrt{1-\zeta^2}\,\omega_n} = \frac{2\pi\zeta}{\sqrt{1-\zeta^2}} \tag{2.120}$$

Squaring Eq. (2.120) and rearranging gives

$$(\delta^2 + 4\pi^2)\zeta^2 = \delta^2 \tag{2.121}$$

and

$$\zeta = \frac{\delta}{\sqrt{\delta^2 + 4\pi^2}} \tag{2.122}$$

Hence, if two peak values x_1 and x_2 are measured, the logarithmic decrement δ can be obtained by using Eq. (2.118), and then damping ratio ζ can be evaluated by using Eq. (2.122). This is a simple, practical way to determine the damping ratio ζ by experiments.

Example 2.6

A mass-spring-damper system decays from 10 cm to 5 cm over one period. Determine its damping ratio.

Solution:

Using Fig. 2.23, it can be found that, $x_1=10$ cm at time t_1 and $x_2=5$ cm at time t_2.

Using logarithmic decrement in Eq. (2.118), we get $\delta=\ln\frac{x_1}{x_2}=\ln\frac{10}{5}=1$.

Using Eq. (2.122), the damping ratio is $\zeta=\frac{\delta}{\sqrt{\delta^2+4\pi^2}}=\frac{1}{\sqrt{1+4\pi^2}}=0.157$.

2.8 Summary: the Effects of Damping on an Unforced Mass-spring System

Then the general solution of Eq. (2.65) depends on the value of damping ratio ζ, and is given by

$$x(t)=\begin{cases}\exp(-\zeta\omega_n t)\left[C_1\cos(\omega_d t)+C_2\sin(\omega_d t)\right] & \zeta<1\\ \exp(-\omega_n t)(C_3+C_4 t) & \zeta=1\\ C_5\exp\left[(-\zeta+\sqrt{\zeta^2-1})\omega_n t\right]+C_6\exp\left[(-\zeta-\sqrt{\zeta^2-1})\omega_n t\right] & \zeta>1\end{cases} \tag{2.123}$$

where C_i, $i=1, 2, 3, 4, 5, 6$, are arbitrary constants, which can be determined by initial conditions.

With the initial conditions $x(0)=x_0$ and $\dot{x}(0)=v_0$. The constants C_i in Eq. (2.123) can be expressed in terms of the initial conditions as follows:

$$C_1=x_0 \tag{2.124a}$$

$$C_2=\frac{v_0+\zeta\omega_n x_0}{\omega_d} \tag{2.124b}$$

$$C_3=x_0 \tag{2.124c}$$

$$C_4=v_0+\omega_n x_0 \tag{2.124d}$$

$$C_5=\frac{v_0+(\zeta+\sqrt{\zeta^2-1})\omega_n x_0}{2\omega_n\sqrt{\zeta^2-1}} \tag{2.124e}$$

$$C_6=\frac{-v_0-(\zeta-\sqrt{\zeta^2-1})\omega_n x_0}{2\omega_n\sqrt{\zeta^2-1}} \tag{2.124f}$$

2.9 MATLAB Examples for Free Vibration of SDOF Systems

In this section, several complete MATLAB examples are given, which can be used to solve the free vibration problems for undamped, underdamping and overdamping SDOF systems, respectively.

2.9.1 Free vibration for undamped SDOF systems

An undamped spring-mass system with natural frequency $\omega_n=2$ rad/s, is subject to an initial displacement $x_0=1$ m, and an initial velocity $v_0=1$ m/s. Plot the time variations of the mass displacement using MATLAB.

In this case, Eqs. (2.13) and (2.18) are used, respectively. The results are shown in Fig. 2.24, it can be found that the same results can be obtained by using Eqs. (2.13) and (2.18). The MATLAB program with comments is as follows.

Line	MATLAB Codes	Comments
1	clearall, close all;	
2	omega_n= 2;	Natural frequency
3	x_0 = 1;	Initial displacement
4	v_0 = 1;	Initial velocity
5	t = 0:0.1:10;	
6	C = sqrt(x_0^2 + (v_0/omega_n)^2);	Amplitude
7	phi = atan2(v_0/omega_n,x_0);	Phase
8	x = C * cos(omega_n * t-phi);	Eq. (2.13)
9	figure(1),	
10	plot(t,x)	
11	hold on	
12	x1=x_0 * cos(omega_n * t)+v_0/omega_n * sin(omega_n * t);	Eq. (2.18)
13	plot(t,x1,'k:','linewidth',2)	
14	xlabel('t')	
15	ylabel('x(t)')	

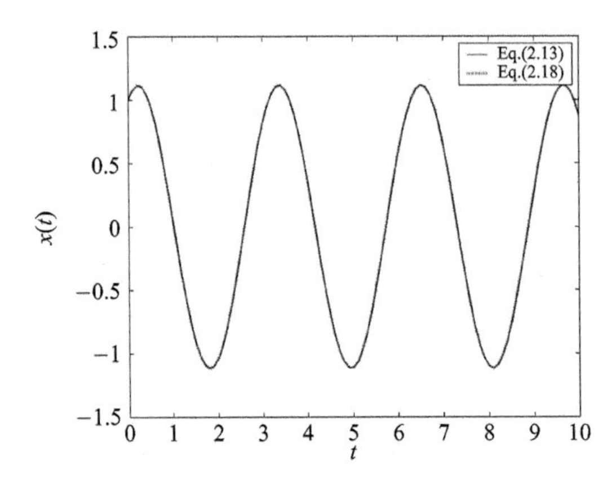

Fig. 2.24 Calculation results by using Eqs. (2.13) and (2.18)

2.9.2 Free vibration for underdamped SDOF systems

An underdamped spring-mass-damper system with mass $m=1$ kg and stiffness $k=10$ N/m, is subject to an initial displacement $x_0=1$ mm, and an initial velocity $v_0=1$ mm/s. Plot the time variations of the mass displacement under different damping using MATLAB.

In this case, Eq. (2.91) can be used to calculate the underdamped system. The results are shown in Fig. 2.25 for damping $c=1$, 2 and 3 Ns/m.

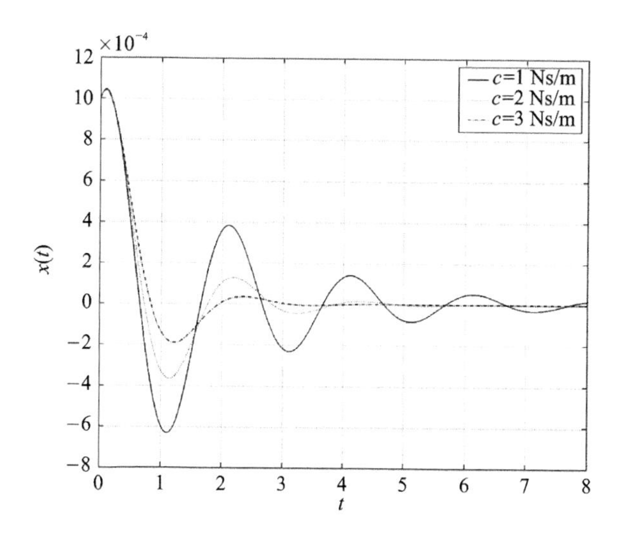

Fig. 2.25 Responses for underdamping SDOF system with different damping

The MATLAB program is as follows.

```
clear all, close all;
m=1;
k=10;
vo=1/1000;
xo=1/1000;
t=linspace(0,8,1000);
wn=sqrt(k/m)
for c= 1: 3
zeta=c/(2 * m * wn)
if zeta>=1 break, end
wd=wn * sqrt(1-zeta^2)
A=(1/wd) * (sqrt(((vo+zeta * wn * xo)^2)+((xo * wd)^2)));
ang=atan((xo * wd)/(vo+zeta * wn * xo));
xt=A * sin(t * wd+ang). * exp(-zeta * t * wn);
display(c)
figure(1);
if c==1 plot(t,xt,'k'), hold on; end
if c==2 plot(t,xt,'k:'), hold on; end
if c==3 plot(t,xt,'k-. '), hold on; end
```

```
end
grid on
ylabel('x(t)')
xlabel ('t')
legend('c=1 Ns/m', 'c=2 Ns/m', 'c=3 Ns/m')
```

2.9.3 Free vibration for critical damped SDOF systems

A critical damped spring-mass-damper system with mass $m=1$ kg and stiffness $k=4$ N/m, is subject to an initial displacement x_0 and an initial velocity v_0. Plot the time variations of the mass displacement under critical damping using MATLAB.

In this case, Eq. (2.105) can be used to calculate the critical damped system. The results are shown in Fig. 2.26 for $x_0=1$ mm, and an initial velocity $v_0=0.01$ m/s, 0 m/s, -0.01 m/s and -0.02 m/s, respectively.

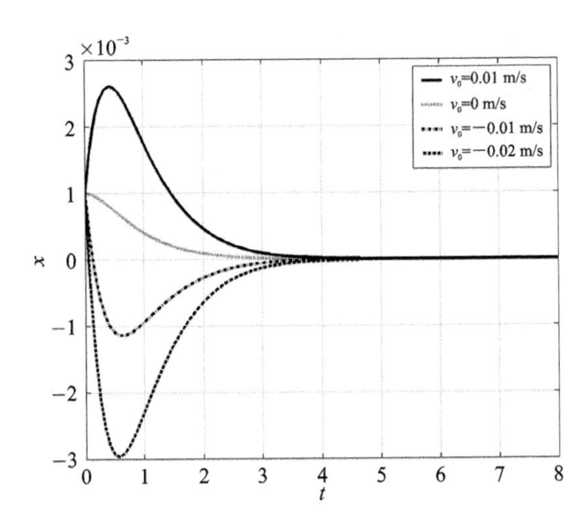

Fig. 2.26 Responses of the critical damping system with different initial conditions

The MATLAB program is as follows.

```
clear all, close all;
% critical-damping case
m=1;
k=4;
t=linspace(0,8,100);
wn=sqrt(k/m);
x0=1e-3;
```

```
for N=1:4
    if N==1 v0=1e-2; end
    if N==2 v0=0; end
    if N==3 v0=-1e-2; end
    if N==4 v0=-2e-2; end
    C1=x0;
    C2=v0+wn * x0;
    xt=C1 * exp(-wn * t) + C2 * t. * exp(-wn * t);
    figure(1);
    if N ==1  plot(t,xt,'k','linewidth',2), end
    hold on,    grid on
    if N ==2  plot(t,xt,'k:','linewidth',2), end
    if N ==3  plot(t,xt,'k-.','linewidth',2), end
    if N ==4  plot(t,xt,'k--','linewidth',2), end
    ylabel('\itx')
    xlabel ('\itt')
end
legend('v_0=0.01m/s','v_0=0m/s','v_0=-0.01m/s','v_0=-0.02m/s')
```

2.9.4 Free vibration for overdamped SDOF systems

An overdamped spring-mass-damper system with mass $m = 100$ kg and stiffness $k = 225$ N/m, is subject to an initial displacement $x_0 = 1$ mm, and an initial velocity $v_0 = 1$ mm/s. Plot the time variations of the mass displacement under different damping using MATLAB.

In this case, Eq. (2.85) can be used to calculate the overdamped system. The results are shown in Fig. 2.27 for damping $c=500$, 600 and 700 Ns/m, respectively.

The MATLAB program is as follows.

```
clear all, close all;
% overdamping
m=100;%kg
k=225;%N/m
x0=1/1000;%m
v0=1/1000;%m/s
t=linspace(0,15,100);
wn=sqrt(k/m)        %natural frequency
```

```
for c=500 :100: 700          %; %kg/s
 zeta=c/(2 * m * wn)           %damping ratio
 a1n= v0+(zeta+(zeta^2-1)^0.5) * wn * x0;
 a2n=-v0-(zeta-(zeta^2-1)^0.5) * wn * x0;
 den=wn * (zeta^2-1)^0.5;
 a1=a1n/(2 * den);
 a2=a2n/(2 * den);
 xt=(a1 * exp(den * t)+a2 * exp(-den * t)). * exp(-zeta * wn * t);
 display(c)
 figure(2);
 plot(t,xt,'k'), hold on
 end
 grid on
 ylabel('x(t)')
 xlabel ('t')
```

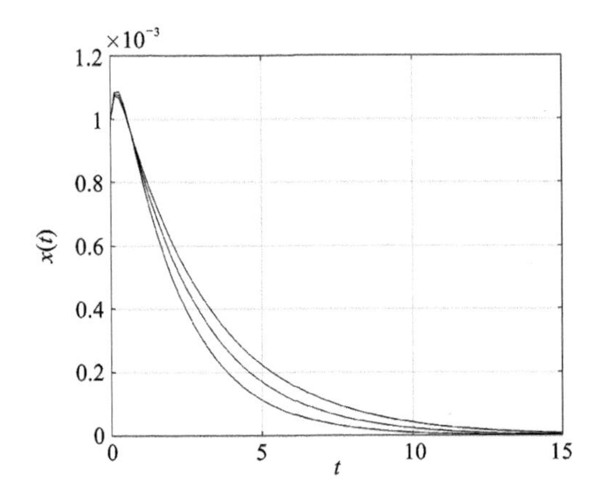

Fig. 2.27　Responses of the overdamped system with different dampers

2.9.5　The GUI program for free vibration of SDOF systems

According to the analysis in above sections, the free vibration problem can be solved by using different MATLAB programs. Furthermore, we offer a graphical user interfaces (GUIs) program based on MATLAB Graphical User Interface Development Environment (GUIDE). In this GUI program, we can input the values of mass, spring and damping. And we also can change the initial displacement and initial velocity as well. The program can automatically identify the system is underdamped, critical damping or overdamped. And the value of the natural frequency as well as the damping ratio are calculated and displayed. Fig. 2. 28 shows the interface of the GUI program.

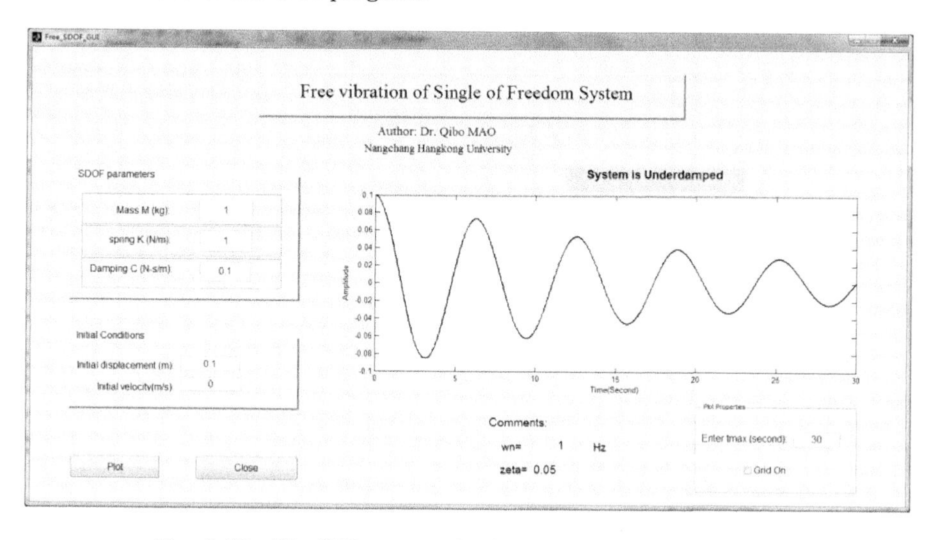

Fig. 2. 28 The GUI program for free vibration of SDOF system

Questions

2. 1 Select the most appropriate answer from the multiple choices given.

(1) Which type of vibrations are also known as transient vibrations?

(A) Undamped vibrations (B) Damped vibrations

(C) Torsional vibrations (D) Transverse vibrations

(2) During transverse vibrations, shaft is subjected to which type of stresses?

(A) Tensile stresses (B) Torsional shear stress

(C) Bending stresses (D) All of the above

(3) Which of the following relations is true when springs are connected in parallel? where K = spring stiffness.

(A) $K_e = K_1 + K_2$ (B) $(1/K_e) = (1/K_1) + (1/K_2)$

(C) $K_e = (1/K_1) + (1/K_2)$ (D) $K_e = K_1 + (1/K_2)$

(4) In which type of vibrations, amplitude of vibration goes on decreasing every cycle?

(A) Damped vibrations (B) Undamped vibrations

(C) Both (a) and (b) (D) None of the above

(5) What are continuous systems?

(A) Systems which have infinite number of degree of freedom

(B) Systems which have finite number of degree of freedom

(C) Systems which have no degree of freedom

(D) None of the above

2.2 Calculate equivalent stiffness of the spring for the system shown in Fig. 2.29, which has spring stiffness of 100 N/m.

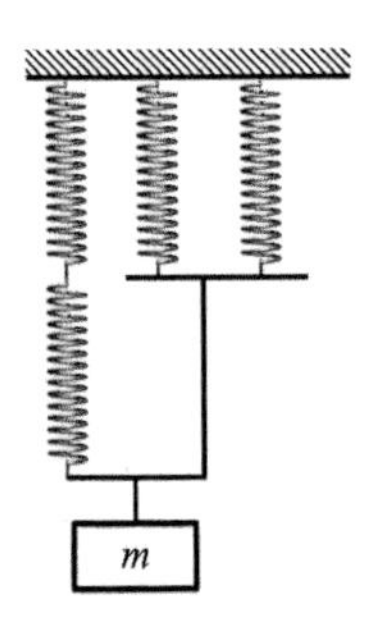

Fig. 2.29 Springs in series and in parallel

2.3 Assume that the stiffness of the springs shown in Fig. 2.30 is k. Determine the equivalent spring constant of the system shown in Fig. 2.30.

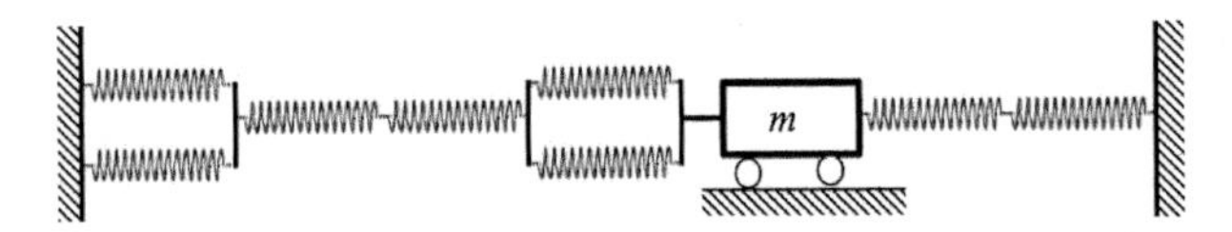

Fig. 2.30 Vibration system with eight springs

2.4 Two springs have spring stiffness of 1 000 N/m and 2 000 N/m respectively. What is the spring stiffness if they are replaced by an equivalent system?

(1) If they are connected in series;

(2) If they are connected in parallel.

2.5 Verify that $x(t)=A\sin(\omega_n t+\phi)$ can be represented as $x(t)=B\sin(\omega_n t)+C\cos(\omega_n t)$ and calculate B and C in terms of A and ϕ.

2.6 Using the solution of $m\ddot{x}+kx=0$ in the form of $x(t)=B\sin(\omega_n t)+C\cos(\omega_n t)$, calculate the values of B and C in terms of initial conditions x_0 and v_0.

2.7 (1) A 0.5 kg mass is attached to a linear spring of stiffness 0.1 N/m. Determine the natural frequency of the system in hertz.

(2) Repeat this calculation for a mass of 50 kg and a stiffness of 10 N/m. Compare the result with that of part (1).

2.8 A vibration system is shown in Fig. 2.31. The mass m_1 moves on a frictionless surface. The pulley is supported on frictionless bearings, and its axis of rotation is fixed. Its moment of inertia about this axis is I. The displacement of the mass is denoted by x and the corresponding rotation of the pulley is denoted by θ. When $x=0$ (and $\theta=0$), the springs k_1 and k_2 are unstretched.

(1) Using Newton's second law, and free-body diagrams, develop an equivalent equation of motion for this system in terms of the response variable x. What is the equivalent mass, and what is the equivalent stiffness of the system?

(2) Verify the result in part (1) using the energy method.

(3) What is the natural frequency of vibration of the system?

(4) Express the equation of the system in terms of the rotational response variable. What is the natural frequency of vibration corresponding to this rotational form of the system equation? What are the equivalent moment of inertia and the equivalent torsional stiffness of the rotational form of the system?

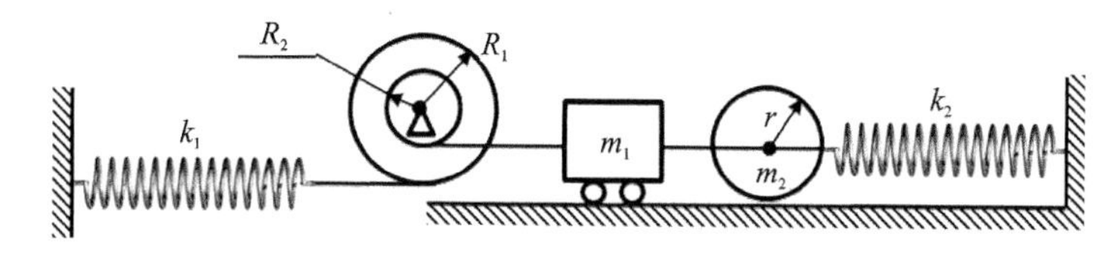

Fig. 2.31 A vibration system

2.9 A uniform heavy spring of mass m_s and stiffness k is attached at one end to a mass m that is free to roll on a frictionless horizontal plane, and the other end is fixed. A schematic diagram of this system is shown in Fig. 2.32. Assume that the velocity distribution along the spring is given by

$$v_s(x) = v\sin\frac{\pi x}{2l}$$

where l is the unstretched length of the spring, v is the velocity of the connected mass, x is the distance of a point along the spring from the fixed end. Determine an equivalent lumped mass of this system to represent the inertia effects of the spring.

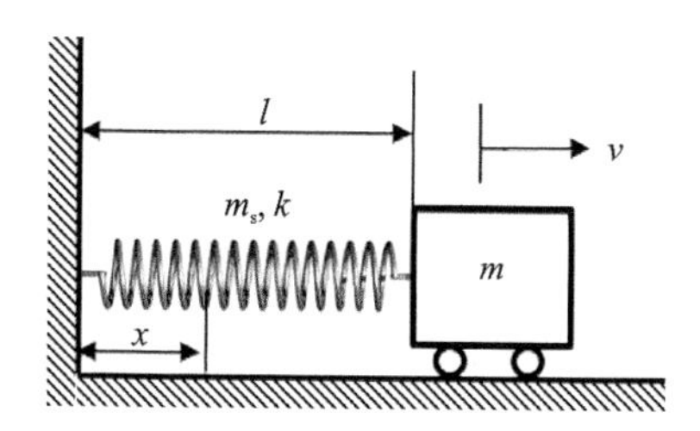

Fig. 2.32 A heavy spring connected to a rolling mass

2.10 A homogeneous cylinder of mass m and radius R is rolling on the horizontal plane with small angle displacement θ, Assume the cylinder rolls without friction and no sliding. There are two horizontal springs with stiffness k, as shown in Fig. 2.33. Find the natural frequency of the system.

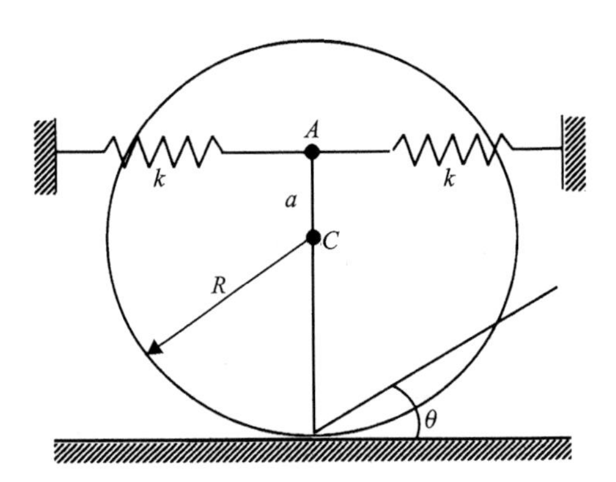

Fig. 2.33 Cylinder rolling on the horizontal plane

2.11 Find the equation of motion for the system of Fig. 2.34, and find the natural frequency. In particular, using static equilibrium along with Newton's law, determine what effect gravity has on the equation of motion and the system's natural frequency. Assume the block slides without friction.

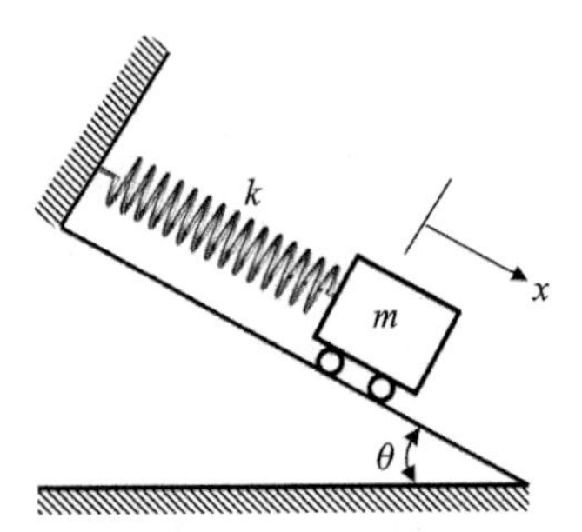

Fig. 2.34 A spring-mass system on frictionless surface

2.12 An undamped system vibrates with a frequency of 10 Hz and amplitude 1 mm. Calculate the maximum amplitude of the system's velocity and acceleration.

2.13 Solve the following initial value problems, and determine the natural frequency, amplitude and phase angle of each solution.

(1) $\ddot{x}+x=0, x(0)=1, \dot{x}(0)=-1$.

(2) $\ddot{x}+25x=0, x(0)=2, \dot{x}(0)=\sqrt{5}$.

(3) $\ddot{x}+100x=0, x(0)=3, \dot{x}(0)=4$.

(4) $\ddot{x}+16x=0, x(0)=4, \dot{x}(0)=5$.

2.14 Solve the following initial value problems. For each problem, determine whether the system is under-, over-, or critically damped.

(1) $\ddot{x}+6\dot{x}+9x=0, x(0)=1, \dot{x}(0)=0$.

(2) $\ddot{x}+4\dot{x}+3x=0, x(0)=0, \dot{x}(0)=1$.

(3) $\ddot{x}+6\dot{x}+10x=0, x(0)=1, \dot{x}(0)=2$.

(4) $\ddot{x}+2\dot{x}+16x=0, x(0)=2, \dot{x}(0)=-2$.

(5) $4\ddot{x}+6\dot{x}+4x=0, x(0)=3, \dot{x}(0)=3$.

(6) $\ddot{x}+8\dot{x}+16x=0, x(0)=-5, \dot{x}(0)=5$.

2.15 Consider a mass-spring system described by the equation $\ddot{x}+2\dot{x}+kx=0$. Give the value(s) of k for which the system is under-, over-, and critically damped.

2.16 Consider a mass-spring system described by the equation $4\ddot{x}+\gamma\dot{x}+100x=0$. Give the value(s) of γ for which the system is under-, over-, and critically damped.

2.17 Derive the solution of $m\ddot{x}+kx=0$ and plot the result for at least two periods for the case with $\omega_n=2$ rad/s, $x_0=2$ mm, and $v_0=\sqrt{5}$ mm/s.

2.18 Solve $m\ddot{x}+kx=0$ for $k=4$ N/m, $m=1$ kg, $x_0=1$ mm, and $v_0=1$ mm/s. Plot the solution.

2.19 The amplitude of vibration of an undamped SDOF system is 1 mm. The phase shift from $t=0$ is 2 rad, and the frequency is 1 rad/s. Calculate the initial conditions that caused this vibration to occur. Assume that the response is of the form $x(t)=A\sin(\omega_n t+\phi)$.

Chapter 3 Harmonic Excitation of SDOF Systems

In Chapter 2, the discussion is concentrated on free vibration problem. It should be noted that the free vibration problem is caused by initial displacement and/or initial velocity and without any external force after the motion has been initiated. Therefore, the free vibration is represented by putting the external force $f(t)=0$ in the dynamic equation of motion.

On the other hand, the forced vibration is the dynamic motion caused by the application of external force (with or without initial displacement and velocity). Therefore, $f(t)\neq 0$ in the equation of motion is for forced vibration problem. Furthermore, forced vibrations have different motion equations for different variations of the applied force with time. In this chapter, the equations for harmonic excitation will be derived analytically for undamped and underdamped SDOF vibration systems.

3.1 Harmonic Excitation

Now let us introduce a nonzero forcing function $f(t)$ into a mass-spring-damper system, as shown in Fig. 3.1. The differential equation of motion, for the system shown in Fig. 3.1, can be expressed as

$$m\ddot{x}+c\dot{x}+kx=f(t) \tag{3.1}$$

where $f(t)$ is the applied force.

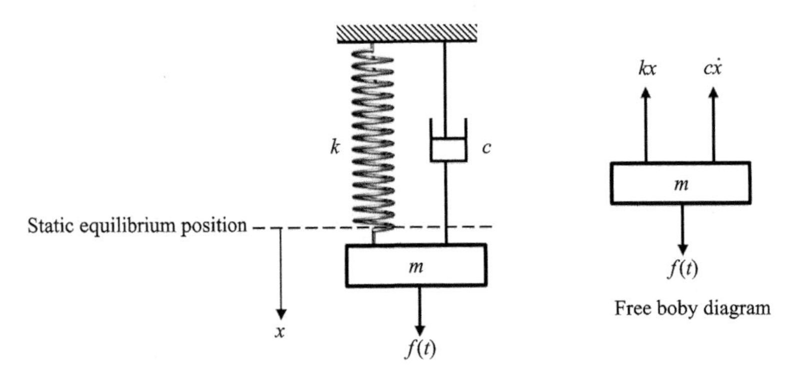

Fig. 3.1 A mass-spring-damper system with an applied force $f(t)$

Assume that

$$f(t)=F_0\sin(\omega t) \tag{3.2}$$

where F_0 is a constant, ω is the excitation frequency. Then the system is termed as a harmonically excited system. Substituting Eq. (3.2) into Eq. (3.1), the differential equation of motion for a harmonically excited system can be expressed as

$$m\ddot{x}+c\dot{x}+kx=F_0\sin(\omega t) \tag{3.3}$$

The next step is to find the general solution of Eq. (3.3). For giving F_0 and ω, the response of the system in Eq. (3.3) can be determined by the initial conditions $x(0)$ and $\dot{x}(0)$.

Assume that x_h is the homogeneous solution of the homogeneous equation problem, such as

$$m\ddot{x}_h+c\dot{x}_h+kx_h=0 \tag{3.4}$$

And assume that x_p is a particular solution of particular equation problem, such as

$$m\ddot{x}_p+c\dot{x}_p+kx_p=F_0\sin(\omega t) \tag{3.5}$$

Then it can be proofed that

$$x(t)=x_h(t)+x_p(t) \tag{3.6}$$

is the general solution of Eq. (3.3).

Proof:

Notice that $x(t)$ is the general solution of Eq. (3.3), on the other hand, by using Eq. (3.6), $x(t)$ can also be expressed in terms of a linear combination of two linearly independent functions. So, it is important to verify that $x(t)$ in Eq. (3.6) satisfies the differential Eq. (3.3).

From Eq. (3.6), the velocity and acceleration can be expressed as

$$\dot{x}(t)=\dot{x}_h(t)+\dot{x}_p(t) \tag{3.7}$$

and

$$\ddot{x}(t)=\ddot{x}_h(t)+\ddot{x}_p(t) \tag{3.8}$$

Substituting Eqs. (3.6), (3.7) and (3.8) into Eq. (3.3), yields

$$\begin{aligned} m\ddot{x}+c\dot{x}+kx &= m(\ddot{x}_h+\ddot{x}_p)+c(\dot{x}_h+\dot{x}_p)+k(x_h+x_p) \\ &= (m\ddot{x}_h+c\dot{x}_h+kx_h)+(m\ddot{x}_p+c\dot{x}_p+kx_p) \\ &= 0+F_0\sin(\omega t) \end{aligned} \tag{3.9}$$

From Eq. (3.9), it can be found that $x(t)$ given by Eq. (3.6) satisfies Eq. (3.3).

The homogeneous solution $x_h(t)$, of the homogeneous problem in Eq. (3.4) has been derived in Chapter 2. And the homogeneous solution $x_h(t)$ can be expressed as

$$x_h(t)=\begin{cases}\exp(-\zeta\omega_n t)\left[C_1\cos(\omega_d t)+C_2\sin(\omega_d t)\right] & \zeta<1\\ \exp(-\omega_n t)(C_3+C_4 t) & \zeta=1\\ C_5\exp\left[(-\zeta+\sqrt{\zeta^2-1})\omega_n t\right]+C_6\exp\left[(-\zeta-\sqrt{\zeta^2-1})\omega_n t\right] & \zeta>1\end{cases} \tag{3.10}$$

Coming next, we should determine a particular solution x_p in Eq. (3.5). Assume that the solution x_p has the following form:

$$x_p(t)=X\sin(\omega t-\phi) \tag{3.11}$$

where X and ϕ are constants, and ω is the frequency of the exciting force (as shown in Eq. (3.2)). Substituting Eq. (3.11) into Eq. (3.5), we get

$$-m\omega^2 X\sin(\omega t-\phi)+c\omega X\cos(\omega t-\phi)+kX\sin(\omega t-\phi)=F_0\sin(\omega t) \tag{3.12}$$

Notice that $\cos\alpha=\sin\left(\alpha+\frac{\pi}{2}\right)$, Eq. (3.12) can be rewritten as

$$-m\omega^2 X\sin(\omega t-\phi)+c\omega X\sin\left(\omega t-\phi+\frac{\pi}{2}\right)+kX\sin(\omega t-\phi)=F_0\sin(\omega t) \tag{3.13}$$

Clearly, Eq. (3.13) is a vector equation, and the graphical interpretation for Eq. (3.13) is shown in Fig. 3.2.

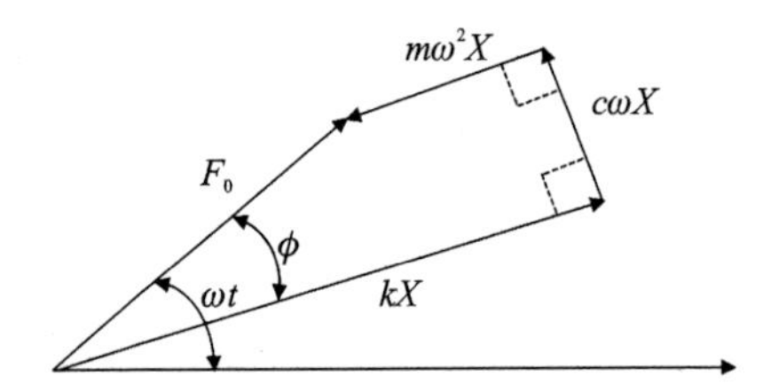

Fig. 3.2 Graphical interpretation for Eq. (3.13)

From Fig. 3.2, F_0 can be expressed as

$$F_0=(kX-m\omega^2 X)^2+(c\omega X)^2 \tag{3.14}$$

From Eq. (3.14), the amplitude X can be written as

$$X=\frac{F_0}{\sqrt{(k-m\omega^2)^2+(c\omega)^2}} \tag{3.15}$$

From Fig. 3.2, we obtain

$$\tan\phi=\frac{c\omega X}{kX-m\omega^2 X}=\frac{c\omega}{k-m\omega^2} \tag{3.16}$$

From Eq. (3.16), the angle ϕ can be written as

$$\phi=\arctan\frac{c\omega}{k-m\omega^2} \tag{3.17}$$

In summary, a particular solution x_p of Eq. (3.5) is given by Eq. (3.11), with constants X and ϕ given in Eqs. (3. 15) and (3. 17), respectively.

3.2 Complex Analysis

Using the fact that the complex exponential is periodic, we also can assume that the harmonically excited force is expressed as exponential form, such as $f(x)=F_0\exp(\mathrm{j}\omega t)$, so the differential equation of motion is rewritten as

$$m\ddot{x}+c\dot{x}+kx=F_0\exp(\mathrm{j}\omega t) \tag{3.18}$$

Notice that $\exp(\mathrm{j}\omega t)=\cos(\omega t)+\mathrm{j}\sin(\omega t)$, so the imaginary component of Eq. (3. 18) is equal to Eq. (3. 3).

We can assume a solution of the form

$$x_{\mathrm{p}}(t)=X\exp[\mathrm{j}(\omega t-\phi)] \tag{3.19}$$

Substituting Eq. (3. 19) into Eq. (3. 18), we get

$$(-m\omega^2+\mathrm{j}c\omega+k)X\exp[\mathrm{j}(\omega t-\phi)]=F_0\exp(\mathrm{j}\omega t) \tag{3.20}$$

Notice that the term exp ($\mathrm{j}\omega t$) in Eq. (3. 20) is general nonzero. Clearly, Eq. (3. 20) can be rewritten as

$$k-m\omega^2+\mathrm{j}c\omega=\frac{F_0}{X}\exp(\mathrm{j}\phi) \tag{3.21}$$

Notice that k, m, c, F_0, X and ω are all real, the graphical interpretation for Eq. (3. 21) can be shown as Fig. 3. 3.

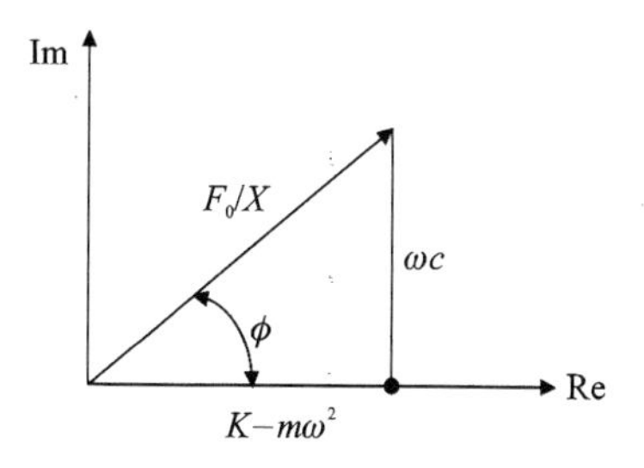

Fig. 3. 3 Graphical interpretation for Eq. (3. 21)

From Fig. 3. 3, it is easy to find that

$$\frac{F_0}{X}=\sqrt{(k-m\omega^2)^2+(c\omega)^2} \tag{3.22}$$

and

$$X=\frac{F_0}{\sqrt{(k-m\omega^2)^2+(c\omega)^2}} \tag{3.23}$$

$$\phi=\arctan\frac{c\omega}{k-m\omega^2} \tag{3.24}$$

Compare Eqs. (3. 23), (3. 24) with Eqs. (3. 15), (3. 17), it can be found that the same solution can be obtained for both cases. Furthermore, if the applied force with cosine form ($f(x)=F_0\cos(\mathrm{j}\omega t)$) can obtain the same solution for the sine case ($f(x)=F_0\sin(\mathrm{j}\omega t)$) or exponential case ($f(x)=F_0\exp(\mathrm{j}\omega t)$). So the applied forces with cosine form, sine form or exponential form are often used interchangeably.

3.3 Undamped SDOF System with Harmonic Excitation

If damping $c=0$, Eq. (3.3) can be rewritten as

$$m\ddot{x}+kx=F_0\sin(\omega t) \tag{3.25}$$

As discussed in preview chapter, it is easy to find that the homogeneous solution for Eq. (3.25) as

$$x_{\mathrm{h}}(t)=A\cos(\omega_{\mathrm{n}}t)+B\sin(\omega_{\mathrm{n}}t) \tag{3.26}$$

where A and B are determined by initial conditions.

3.3.1 Excitation frequency ≠ natural frequency

In this case, using Eqs. (3.11), (3.15) and (3.17), the particular solution x_{p} of Eq. (3.25) can be written as

$$x_{\mathrm{p}}(t)=\frac{F_0}{k-m\omega^2}\sin(\omega t)\quad \text{with}\quad \omega\neq\omega_{\mathrm{n}} \tag{3.27}$$

Define the frequency ratio $\beta=\frac{\omega}{\omega_{\mathrm{n}}}$, which is the ratio of the applied loading frequency ω to the natural frequency ω_{n}. Notice that $k=m\omega_{\mathrm{n}}^2$. Then Eq. (3.27) can be rewritten as

$$x_{\mathrm{p}}(t)=\frac{F_0}{k(1-\beta^2)}\sin(\omega t) \tag{3.28}$$

So the general solution of Eq. (3.25) can be represented as

$$x(t)=x_{\mathrm{h}}(t)+x_{\mathrm{p}}(t)=\underbrace{A\cos(\omega_{\mathrm{n}}t)+B\sin(\omega_{\mathrm{n}}t)}_{\text{Solution for Homogenous equation}}+\underbrace{\frac{F_0}{k(1-\beta^2)}\sin(\omega t)}_{\text{Solution for particular equation}} \tag{3.29}$$

Now the unknown parameters A and B in Eq. (3.29) can be determined by using the initial conditions (such as initial displacement $x(0)=x_0$ and initial velocity $\dot{x}(0)=v_0$). It is easy to find that

$$x(0)=A=x_0 \tag{3.30}$$

$$\dot{x}(0)=B\omega_{\mathrm{n}}+\frac{F_0\omega}{k(1-\beta^2)}=v_0 \tag{3.31a}$$

and

$$B = \frac{v_0}{\omega_n} - \frac{\omega}{\omega_n}\frac{F_0}{k(1-\beta^2)} = \frac{v_0}{\omega_n} - \frac{F_0\beta}{k(1-\beta^2)} \tag{3.31b}$$

Substituting Eqs. (3.30) and (3.31b) into Eq. (3.29), the response of Eq. (3.25) can be obtained as

$$x(t) = x_0\cos(\omega_n t) + \left[\frac{v_0}{\omega_n} - \frac{F_0\beta}{k(1-\beta^2)}\right]\sin(\omega_n t) + \frac{F_0}{k(1-\beta^2)}\sin(\omega t)$$

$$= \underbrace{x_0\cos(\omega_n t) + \frac{v_0}{\omega_n}\sin(\omega_n t)}_{\text{Free response}} + \underbrace{\frac{F_0}{k(1-\beta^2)}[\sin(\omega t) - \beta\sin(\omega_n t)]}_{\text{Forced response}} \tag{3.32}$$

where $\frac{1}{1-\beta^2}$ is termed as the Magnification Factor (MF) representing the amplification effect of the harmonically applied loading.

If $x(0) = \dot{x}(0) = 0$ (zero initial condition), Eq. (3.32) can be simplified as

$$x(t) = \frac{F_0}{k(1-\beta^2)}[\sin(\omega t) - \beta\sin(\omega_n t)] \tag{3.33}$$

In Eq. (3.33), $\sin(\omega t)$ represents the response component at the excitation frequency, it is termed as the steady-state response and is directly related to the excitation force. Also $\beta\sin(\omega_n t)$ is the response component at the natural frequency and is the free vibration effect controlled by the initial conditions. Since in a practical case, damping will cause the last term to vanish eventually, it is termed the transient response. For this hypothetical undamped system, however, this term will not damp out but will continue indefinitely.

In Eq. (3.33), if the natural frequency and the excitation frequency are close but not equal, a beat phenomenon will occur. In this kind of vibration, the amplitude builds up and then diminishes in a regular pattern.

Assume that $k=1$, $F_0=1$, and $\omega_n=1$ in Eq. (3.33). Fig. 3.4 shows the beat phenomenon under different frequency ratios β.

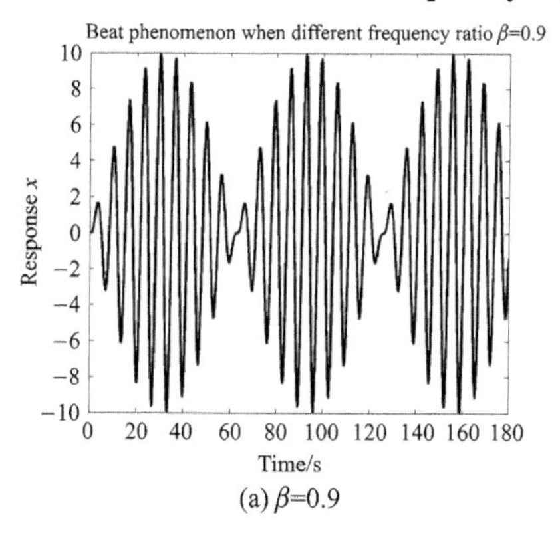

(a) β=0.9

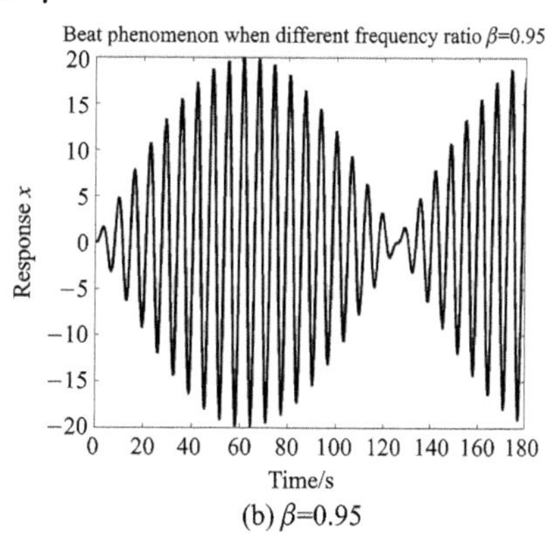

(b) β=0.95

Fig. 3.4 The beat phenomenon under different frequency ratios β

3.3.2 Excitation frequency = natural frequency (resonant condition)

In Eq. (3.32), it can be found that $1-\beta^2=0$ if the excitation frequency $\omega=\omega_n=\sqrt{\frac{k}{m}}$, and then $x(t)$ in Eq. (3.32) becomes infinite. This is the resonance phenomenon. When the excitation frequency equals the undamped natural vibration frequency, is called resonance.

Notice that we cannot directly obtain the solution by using Eq. (3.32) when $\omega=\omega_n$. To solve this problem, we apply L'Hospital's rule to evaluate the limit of the last term in Eq. (3.32),

$$\lim_{\omega\to\omega_n}\left\{\frac{F_0}{k(1-\beta^2)}\left[\sin(\omega t)-\beta\sin(\omega_n t)\right]\right\}$$

$$=\frac{F_0}{k}\lim_{\omega\to\omega_n}\left[\frac{\sin(\omega t)-\beta\sin(\omega_n t)}{1-\frac{\omega^2}{\omega_n^2}}\right]=\frac{F_0}{k}\lim_{\omega\to\omega_n}\left\{\frac{\frac{d}{d\omega}\left[\sin(\omega t)-\beta\sin(\omega_n t)\right]}{\frac{d}{d\omega}\left(1-\frac{\omega^2}{\omega_n^2}\right)}\right\}$$

$$=\frac{F_0}{k}\lim_{\omega\to\omega_n}\left[\frac{t\cos(\omega t)}{-2\frac{\omega}{\omega_n^2}}\right]=\frac{-\omega_n F_0}{2k}t\cos(\omega_n t) \tag{3.34}$$

According to above analysis, it is easy to find that the complete solution when the excitation frequency is equal to natural frequency ($\omega=\omega_n$),

$$x(t)=x_0\cos(\omega_n t)+\frac{v_0}{\omega_n}\sin(\omega_n t)-\frac{\omega_n F_0}{2k}t\cos(\omega_n t) \tag{3.35}$$

From Eq. (3.35), it can be found that the last term $\frac{\omega_n F_0}{2k}t\cos(\omega_n t)$ will tend to infinite as time t goes to infinite, so Eq. (3.35) is an unstable response because the forced response increases steadily. Notice that the same undamped vibration system gives a bounded response for some excitations. However, it could produce an unstable (steady linear increase) response when the excitation frequency is equal to its natural frequency. Hence, the system is not quite unstable, but is not quite stable either. In fact, the undamped oscillator is said to be marginally stable.

To avoid resonance, we should introduce damping and avoid harmonic excitation with frequency equal to the natural frequency of the system. Technically, true resonance only occurs if all of the conditions below are satisfied:

(1) There is no damping: $c=0$;

(2) A periodic forcing function is present;

(3) The frequency of the forcing function exactly matches the natural frequency of the vibration system.

However, it should be notitced that the similar behaviors (unexpectedly large amplitude of oscillation due to a fairly low-strength forcing function) can occur if the damping is very small.

3.3.3 Response ratio for undamped SDOF system

A convenient measure of the influence of dynamic loading is provided by the ratio of the dynamic displacement response to the displacement produced by static application of load F_0, such as,

$$R(t)=\frac{x(t)}{x_{st}}=\frac{x(t)k}{F_0} \tag{3.36}$$

where $x_{st}=\frac{F_0}{k}$ is the displacement which would be produced by the load F_0 applied statically. $R(t)$ is defined as response ratio.

Substituting Eq. (3.33) into Eq. (3.36), it is clearly that the response ratio R of an undamped system can be expressed as

$$R(t)=\frac{1}{1-\beta^2}[\sin(\omega t)-\beta\sin(\omega_n t)] \tag{3.37}$$

It is informative to examine this response behavior in more detail by reference to Fig. 3.5. Fig. 3.5(a) represents the steady-state component of response, it is $R_s(t)=\frac{1}{1-\beta^2}\sin(\omega t)$, while Fig. 3.5(b) represents the so-called transient response, it is $R_t(t)=\frac{1}{1-\beta^2}[-\beta\sin(\omega_n t)]$. The total response ratio $R(t)$, i. e., the sum of both types of response (see Eq. (3.37)), is shown in Fig. 3.5(c). In this example, it is assumed that the frequency ratio $\beta=0.75$, that is, the applied excitation frequency is 0.75 of the natural frequency of the SDOF system. From Fig. 3.5, two conclusions can be obtained:

(1) The tendency for the two components to get in phase and then out of phase again, causing a beating effect in the total response, it can be seen more clearly if the frequency ratio is close to one, as shown in Fig. 3.6.

(2) Notice that it is starting from at rest initial conditions, it should satisfies the specified initial condition $x(0)=\dot{x}(0)=0$. It means that the initial velocity of the transient response can cancel the initial velocity of the steady-state response, so the total response ratio has zero slope at time $t=0$.

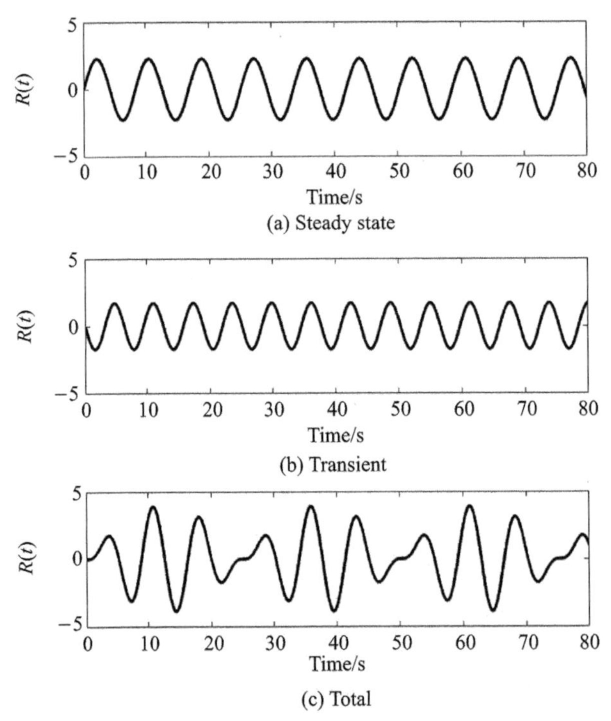

Fig. 3. 5 Response ratio produced by sine wave excitation starting from at rest initial conditions for undamped system when frequency ratio $\beta = 0.75$

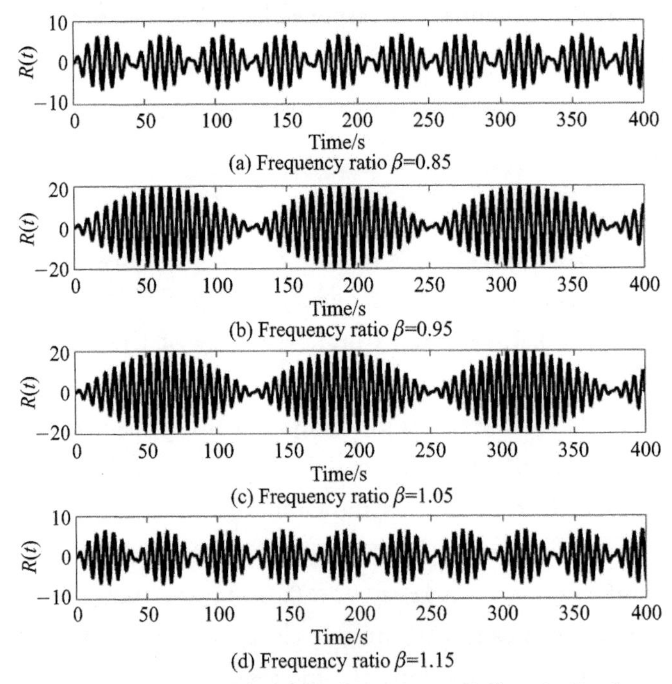

Fig. 3. 6 Response ratio produced by sine wave excitation starting from at rest initial conditions for undamped system with different frequency ratio β

3.4 Damped SDOF System with Harmonic Excitation

3.4.1 Response for damped SDOF system with harmonic excitation

Returning to the equation of motion including viscous damping (see Eq. (3.3)), the response can be summarized as

$$x(t)=\begin{cases}\exp(-\zeta\omega_n t)\left[C_1\cos(\omega_d t)+C_2\sin(\omega_d t)\right]+X\sin(\omega t-\phi) & \zeta<1\\ \exp(-\omega_n t)(C_3+C_4 t)+X\sin(\omega t-\phi) & \zeta=1\\ C_5\exp\left[(-\zeta+\sqrt{\zeta^2-1})\omega_n t\right]+C_6\exp\left[(-\zeta-\sqrt{\zeta^2-1})\omega_n t\right]+X\sin(\omega t-\phi) & \zeta>1\end{cases} \tag{3.38}$$

Fig. 3.7 shows the typical damped vibration response. The general solution in Eq. (3.38) shows that for $\zeta>0$ the homogeneous part of the solution vanishes as $t\rightarrow\infty$ (see Fig. 3.7(a)). We therefore say that the particular solution x_p, given by Eq. (3.11), is the steady state harmonic response of the system (see Fig. 3.7(b)). The total response is the sum of the transient response and the steady state harmonic response, as shown in Fig. 3.7(c).

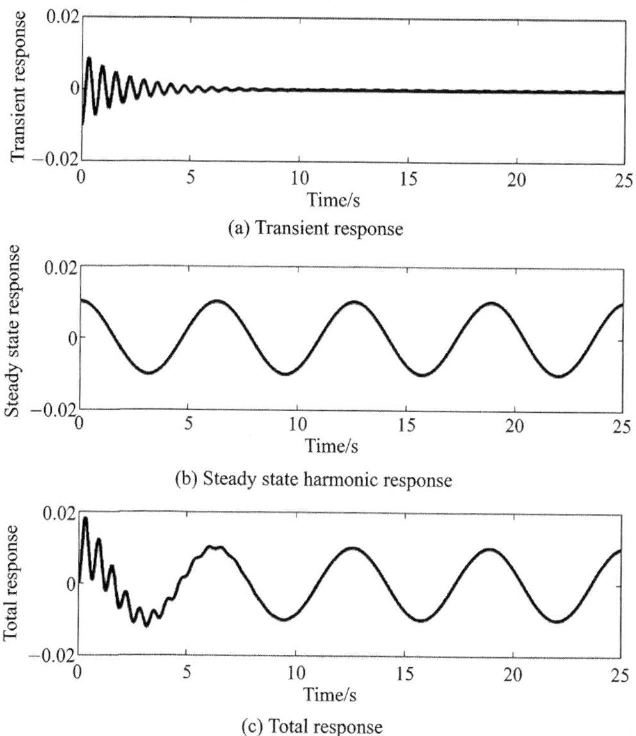

(a) Transient response

(b) Steady state harmonic response

(c) Total response

Fig. 3.7 Typical damped vibration response

3.4.2 Dynamic magnification factor for damped SDOF system

By using frequency ratio $\beta=\frac{\omega}{\omega_n}$ and damping ratio $\zeta=\frac{c}{2\sqrt{km}}$, the amplitude X and angle ϕ for solution $x_p(t)=X\sin(\omega t-\phi)$ (as shown in Eq. (3.11)) can be rewritten as

$$X=\frac{F_0}{k\sqrt{(1-\beta^2)^2+(2\zeta\beta)^2}} \tag{3.39}$$

$$\phi=\arctan\frac{2\zeta\beta}{1-\beta^2} \tag{3.40}$$

The ratio of the resultant harmonic response amplitude to the static displacement which would be produced by the force F_0 will be called the dynamic magnification factor D, such as

$$D=\frac{X}{\frac{F_0}{k}}=\frac{X}{x_{st}}=\frac{1}{\sqrt{(1-\beta^2)^2+(2\zeta\beta)^2}} \tag{3.41}$$

It is seen that both the dynamic magnification factor D and the phase angle ϕ vary with the frequency ratio β and damping ratio ζ, as shown in Figs. 3.8 and 3.9. The following MATLAB program can be used to calculate Eq. (3.41), and the results are shown in Fig. 3.8.

MATLAB Codes for Fig. 3.8	Comments
clear all, close all	Remove all variables in workspace and close all figures. It is a good way to start a new MATLAB script (not function)
beta=linspace(0,3,1e3);	Generates a row vector beta (frequency ratio) of 1 000 points linearly spaced from 0 to 3
damp=[0.0 0.2 0.4 0.6 0.8];	Define damping ratios
for m=1:5	
D1=1./sqrt((1-beta.^2).^2+(2*damp(m)*beta).^2);	Calculate the dynamic magnification factor D by using Eq. (3.31)
if m==1 plot(beta, D1,'k'), hold on, end	Plot the results
if m==2 plot(beta, D1,'k:'), end	
if m==3 plot(beta, D1,'k-.'), end	
if m==4 plot(beta, D1,'k--'), end	
if m==5 plot(beta, D1,'k','linewidth',2), end	
end	

(continued)

ylim([0 3])	When damping ratio = 0, the dynamic magnification factor D will tend to infinite. To display the results clearly, we set y-axis limits in [0,3]
xlabel('\it\beta')	
ylabel('\itD')	
legend('damping ratio = 0', 'damping ratio = 0.2', 'damping ratio = 0.4', ... 'damping ratio = 0.6', 'damping ratio = 0.8')	Display a legend on the figure

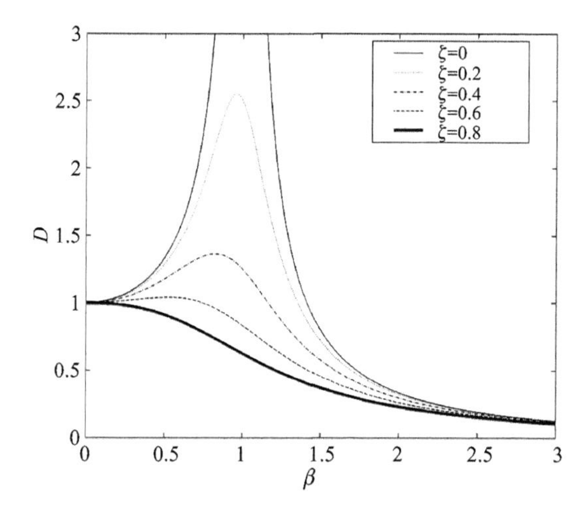

Fig. 3.8 The dynamic magnification factor D varies with the frequency ratio β and damping ratio ζ

The following MATLAB program can be used to calculate the phase of the dynamic magnification factor D, the results are shown in Fig. 3.9. Due to this MATLAB program is similar to program for Fig. 3.8. So there is no comment added in this program.

```
% For Fig. 3.9
clear all, close all
beta=linspace(0,3,1e3);
damp=[0 0.2 0.4 0.6 0.8];
for m=1:5
    D1=atan2(2 * damp(m) * beta,(1-beta.^2)) * 180/pi;
    if m==1 plot(beta, D1,'k'), hold on, end
    if m==2 plot(beta, D1,'k:'),   end
```

```
    if m==3 plot(beta, D1,'k-.'),   end
    if m==4 plot(beta, D1,'k--'),   end
    if m==5 plot(beta, D1,'k','linewidth',2),    end
end
ylim([-5 185])
xlabel('\it\beta')
ylabel('angle (Deg)')
legend('damping ratio = 0', 'damping ratio = 0.2', 'damping ratio= 0.4',...
'damping ratio = 0.6', 'damping ratio = 0.8',4)
```

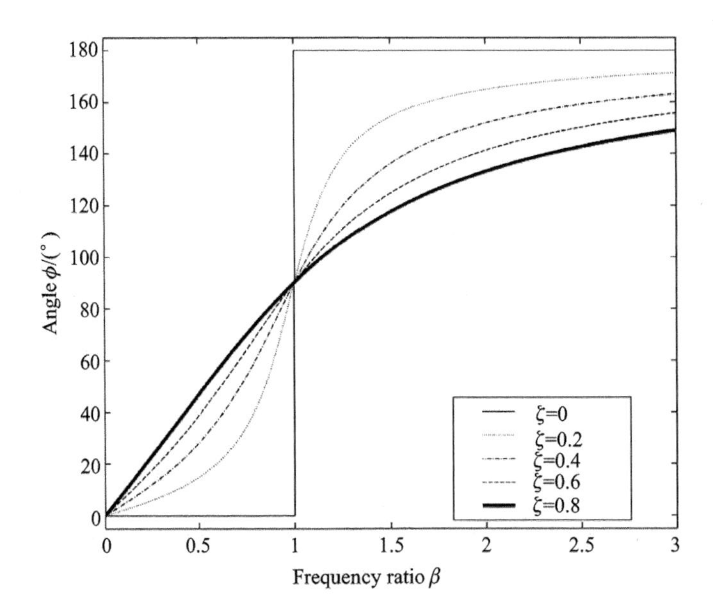

Fig. 3.9 The phase angle ϕ varies with the frequency ratio β and damping ratio ζ

From Eq. (3.41), it is seen that the dynamic magnification factor for $\beta = 1$ (resonant response) can be expressed as

$$D_{\beta=1} = \frac{1}{2\zeta} \tag{3.42}$$

To find the maximum or peak value of dynamic magnification factor, one must differentiate Eq. (3.41) with respect to β and solve the resulting expression for β obtaining

$$\beta_{\text{peak}} = \sqrt{1-2\zeta^2} \tag{3.43}$$

Clearly, for damping ratios $\zeta < \sqrt{2}$, β_{peak} in Eq. (3.43) is the positive real value and then substitute this value of frequency ratio back into Eq. (3.41) giving

$$D_{\text{peak}}=\frac{1}{2\zeta\sqrt{1-\zeta^2}}=\frac{1}{2\zeta}\cdot\frac{\omega_n}{\omega_d} \tag{3.44}$$

For typical values of structural damping ratio, say $\zeta<0.10$, the difference between Eq. (3.44) and the simpler Eq. (3.42) is very small, the difference being 0.13% for $\zeta=0.05$ and 0.5% for $\zeta=0.10$. It means that we can estimate the peak value of dynamic magnification factor by setting $\beta=1$ for low damping ratio cases.

For a more complete understanding of the nature of the resonant response of a structure to harmonic loading, it is necessary to consider the general response Eq. (3.38) for $\zeta<1$, which includes the transient term as well as the steady-state term. At the resonant exciting frequency ($\beta=1$), this equation can be rewritten as

$$x(t)=\exp(-\zeta\omega_n t)\left[C_1\cos(\omega_d t)+C_2\sin(\omega_d t)\right]+X\sin(\omega_n t-\varphi)\quad(\zeta<1) \tag{3.45}$$

Notice that at $\beta=1$, it is easy to find that $\omega=\omega_n$, $\phi=90°$, and $X=\frac{F_0}{k}D=\frac{F_0}{2\zeta k}$, so Eq. (3.45) can be rewritten as

$$x(t)=\exp(-\zeta\omega_n t)\left[C_1\cos(\omega_d t)+C_2\sin(\omega_d t)\right]-\frac{F_0}{2\zeta k}\cos(\omega_n t) \tag{3.46}$$

Assuming that the system starts from rest (such as $x(0)=\dot{x}(0)=0$). The constants C_1 and C_2 can be expressed as follows:

$$C_1=\frac{F_0}{2\zeta k}\quad\text{and}\quad C_2=\frac{F_0}{k}\cdot\frac{\omega_n}{2\omega_d}=\frac{F_0}{k}\cdot\frac{1}{2\sqrt{1-\zeta^2}} \tag{3.47}$$

Substituting Eq. (3.47) into Eq. (3.46), we get

$$x(t)=\frac{F_0}{2\zeta k}\left\{\exp(-\zeta\omega_n t)\left[\cos(\omega_d t)+\frac{\zeta}{\sqrt{1-\zeta^2}}\sin(\omega_d t)\right]-\cos(\omega_n t)\right\} \tag{3.48}$$

3.4.3 Response ratio for $\beta=1$

If the damping ratios in structural systems are small, the term $\sqrt{1-\zeta^2}$ is nearly equal to unity and $\omega_d\approx\omega_n$. In this case, Eq. (3.48) can be rewritten in the approximate form:

$$\begin{aligned}x(t)&\approx\frac{F_0}{2\zeta k}\{\exp(-\zeta\omega_n t)\left[\cos(\omega_n t)+\zeta\sin(\omega_n t)\right]-\cos(\omega_n t)\}\\&=\frac{F_0}{2\zeta k}\{\left[\exp(-\zeta\omega_n t)-1\right]\cos(\omega_n t)+\exp(-\zeta\omega_n t)\cdot\zeta\sin(\omega_n t)\}\end{aligned} \tag{3.49}$$

Recall Eq. (3.36), the response ratio $R(t)$ is defined as

$$R(t)=\frac{x(t)}{\frac{F_0}{k}}\approx\frac{1}{2\zeta}\{[\exp(-\zeta\omega_n t)-1]\cos(\omega_n t)+\exp(-\zeta\omega_n t)\cdot\zeta\sin(\omega_n t)\}\tag{3.50}$$

For zero damping ratio ($\zeta = 0$), this approximate equation is indeterminate; but when L'Hospital's rule is applied, the response ratio for the undamped system can be written as

$$R(t)\approx\frac{1}{2}[\sin(\omega_n t)-\omega_n t\cdot\cos(\omega_n t)]\tag{3.51}$$

Assume that $\omega_n=2$ rad/s, by using Eqs. (3.50) and (3.51), the response ratios for undamped and damped system are shown in Fig. 3.10. Notice that the terms containing $\sin(\omega t)$ contribute little to the response in Eq. (3.51), the peak values in Fig. 3.10(a) build up linearly for the undamped case, changing by an amount π in each cycle. However, they build up in accordance with $\frac{\exp(-\zeta\omega_n t)-1}{2\zeta}$ for the damped case, as shown in Fig. 3.10(b).

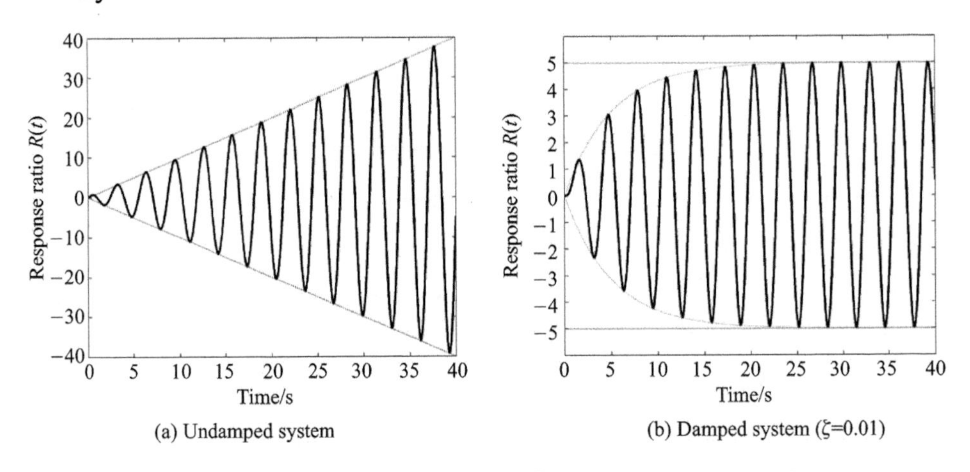

(a) Undamped system (b) Damped system (ζ=0.01)

Fig. 3.10 Response to resonant loading $\beta = 1$ for zero initial conditions

To further show the damping ratio effect on the response ratio. Fig. 3.11 shows the envelope function $\frac{\exp(-\zeta\omega_n t)-1}{2\zeta}$ under different values of damping ratio. The same results are shown in Fig. 3.11(a) and (b) by using different coordinates. From Fig. 3.11(a), it can be found that envelope function is reduced by increasing the damping ratios, as expected. From Fig. 3.11(b), it can be found that the envelope function toward the steady-state level $\frac{1}{2\zeta}$ increases with damping ratio. From both figures, it can be found that the envelope function to nearly steady-state level for higher damping ratio cases is more quickly than that for lower damping ratio cases. For example, if damping ratio is 0.2, the response tends to the steady-state level after 2 seconds ($\omega t = 4\pi$ in Fig. 3.11(b)). However, for a case with damping ratio 0.05, the response tends to the steady-state level after 7 seconds.

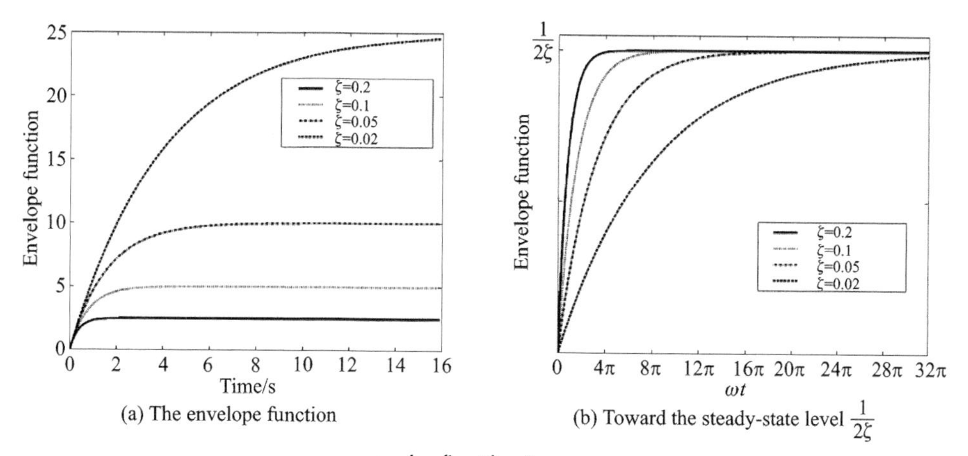

Fig. 3.11 The envelope function $\frac{\exp(-\zeta\omega_n t)-1}{2\zeta}$ under different values of damping ratio

3.5 Harmonic Base Excitation

Assume that a mass-spring-damper system is excited by base motion, as shown in Fig. 3.12. Assume that the base moves according to the function

$$y(t)=Y\sin(\omega t) \tag{3.52}$$

where Y is the constant amplitude of the base motion.

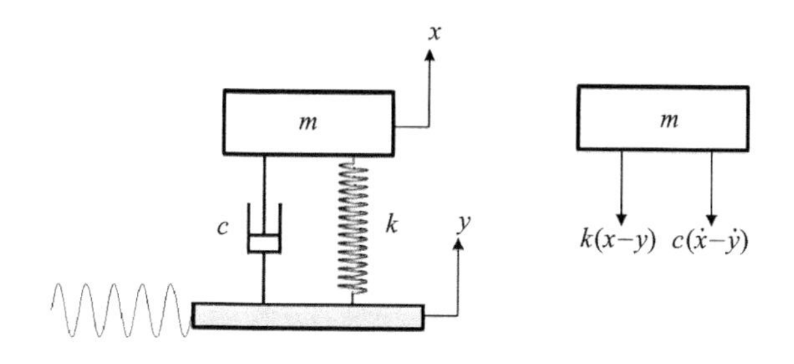

Fig. 3.12 Harmonic base excitation

Assume that $x(t)$ is the displacement of the mass m from its static equilibrium position. So the spring force can be written as $k(x-y)$, the damper force is $c(\dot{x}-\dot{y})$, and the differential equation of motion can be expressed as

$$m\ddot{x}+c(\dot{x}-\dot{y})+k(x-y)=0 \tag{3.53}$$

where m is the mass of the system, c is the viscous damping coefficient, and k is the linear spring stiffness.

3.5.1 Relative motion

Sometimes we are concerned with the relative motion of the mass with respect to the base (such as accelerometer and the velocity meter). In this case, we can define $z=x-y$. The equation of motion in Eq. (3.53) can be rewritten as

$$m\ddot{z}+c\dot{z}+kz=-m\ddot{y} \tag{3.54}$$

From Eq. (3.52), we have

$$\ddot{y}(t)=-\omega^2 Y\sin(\omega t) \tag{3.55}$$

Substituting Eq. (3.55) into Eq. (3.54), yields

$$m\ddot{z}+c\dot{z}+kz=mY\omega^2\sin(\omega t) \tag{3.56}$$

Compare Eq. (3.5) with Eq. (3.56), it can be found that Eq. (3.56) is identical to the previous case (harmonic force excitation). Similar to Eq. (3.5), with F_0 replaced by $mY\omega^2$. The steady-state solution for Eq. (3.56) can be written as

$$z(t)=Z\sin(\omega t-\phi) \tag{3.57}$$

and the steady-state amplitude for this case is

$$Z=\frac{mY\omega^2}{\sqrt{(k-m\omega^2)^2+(c\omega)^2}} \tag{3.58}$$

and the phase lag between the motion of the base and the mass can be written as

$$\phi=\arctan\frac{c\omega}{k-m\omega^2}$$

By using frequency ratio $\beta=\frac{\omega}{\omega_n}$ and damping ratio $\zeta=\frac{c}{2\sqrt{km}}$, it is useful to rewritten Eq. (3.58) into dimensionless form:

$$\frac{Z}{Y}=\frac{m\omega^2}{\sqrt{(k-m\omega^2)^2+(c\omega)^2}}=\frac{\beta^2}{\sqrt{(1-\beta^2)^2+(2\zeta\beta)^2}} \tag{3.59}$$

and

$$\phi=\arctan\frac{2\zeta\beta}{1-\beta^2} \tag{3.60}$$

The variation of the steady-state amplitude Z can be plotted as a function of the forcing frequency. It is convenient to plot in dimensionless, so they become independent of a system's parameters. This is accomplished by plotting Eq. (3.59) as Z/Y versus β in Fig. 3.13, where $\beta=\omega/\omega_n$, with ω_n is constant. The functions are evaluated for various values of damping ratio ζ. The associated phase angles ϕ and β are also shown in Fig. 3.13(b). Notice that the MATLAB program for this case is similar to Figs. 3.8 and 3.9, so the reader can easily modify these MATLAB programs to plot the results shown in Fig. 3.13.

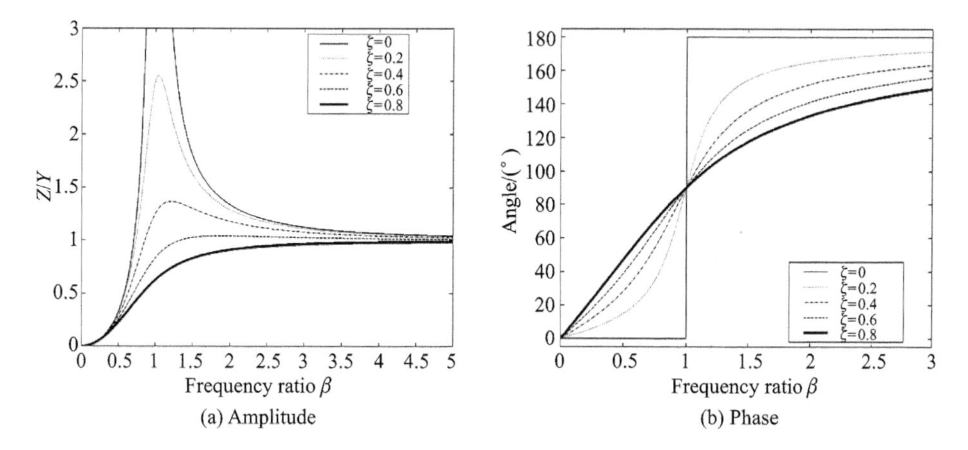

Fig. 3.13 The relative motion Z/Y versus frequency ratio β for harmonic base excitation

Application: vibration measuring instruments

For negligible damping ($\zeta \ll 1$), Eq. (3.59) can be simplified as

$$\frac{Z}{Y}=\frac{\beta^2}{|1-\beta^2|} \tag{3.61}$$

Case 1: $\beta \ll 1$ so $1-\beta^2=1$ in Eq. (3.61), and $Z=Y\beta^2=Y\frac{\omega^2}{\omega_n^2}$.

At low frequency ranges, Z is proportional to $\omega^2 Y$, it means that Z is proportional to acceleration. So in this range the accelerometer works, as shown in Fig. 3.14(a). It must have a high ω_n to extend the available frequency range for accelerometer.

Case 2: $\beta \gg 1$ so $Z/Y=1$.

At high frequency ranges, the ratio between Z and Y is one, that is, Z is equal to the displacement, as shown in Fig. 3.14(b). In this range the vibrometer (which is used to measure the displacement) works. It must have a low ω_n for vibrometer to measure the displacement of low frequency vibrations.

3.5.2 Absolute motion

If we are concerned with the absolute motion, the equation of motion in Eq. (3.53) can be rewritten as

$$m\ddot{x}+c\dot{x}+kx=ky+c\dot{y} \tag{3.62}$$

Here it becomes more convenient to assume the base motion having the exponential form, such as

$$y=Y\exp(\mathrm{j}\omega t) \tag{3.63}$$

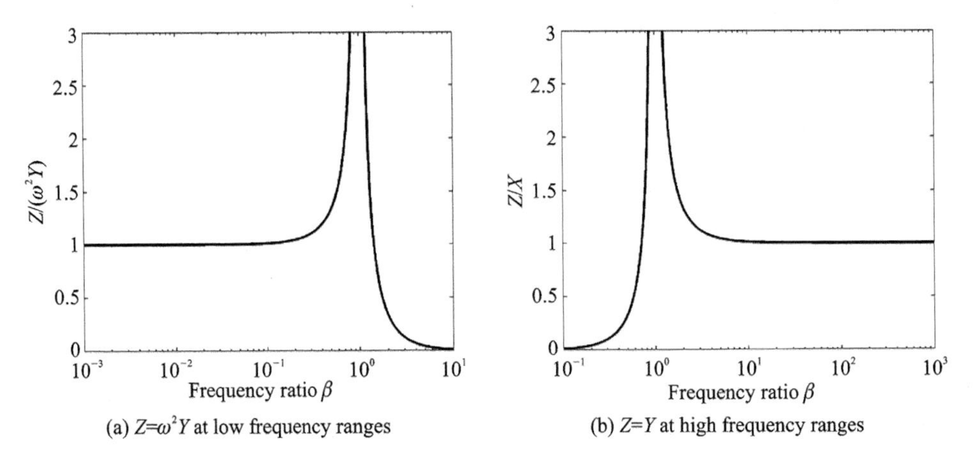

(a) $Z=\omega^2 Y$ at low frequency ranges

(b) $Z=Y$ at high frequency ranges

Fig. 3.14 Case 1 and Case 2

Similar to Section 3.2, the solution for Eq. (3.62) can be expressed as

$$x = X\exp[\mathrm{j}(\omega t - \phi)] \tag{3.64}$$

Substituting Eq. (3.64) into the equation of motion (3.62), yields

$$(-m\omega^2 + \mathrm{j}\omega c + k)X\exp[\mathrm{j}(\omega t - \phi)] = (k + \mathrm{j}\omega c)Y\exp(\mathrm{j}\omega t) \tag{3.65}$$

From Eq. (3.65), it is easy to found that

$$\frac{X}{Y}\exp(-\mathrm{j}\phi) = \frac{k + \mathrm{j}\omega c}{-m\omega^2 + \mathrm{j}\omega c + k} = \frac{k + \mathrm{j}\omega c}{k - m\omega^2 + \mathrm{j}\omega c} \tag{3.66}$$

Eq. (3.66) can be rewritten into dimensionless form, we get

$$\left|\frac{X}{Y}\right| = \frac{\sqrt{1 + (2\zeta\beta)^2}}{\sqrt{(1-\beta^2)^2 + (2\zeta\beta)^2}} \tag{3.67}$$

and

$$\phi = \arctan\frac{2\zeta\beta^3}{1 - \beta^2 + (2\zeta\beta)^2} \tag{3.68}$$

It is seen that both the amplitude $|X/Y|$ and the phase angle ϕ vary with the frequency ratio β and damping ratio ζ, as shown in Figs. 3.15 and 3.16. The following MATLAB program can be used to plot Fig. 3.15.

```
% For Fig. 3.15
clear all, close all
beta=linspace(0,5,1e3);
damp=[0.0 0.2 0.4 0.6 0.8 1];
for m=1:5
    D1=sqrt(1+(2 * damp(m) * beta).^2)./sqrt((1-beta.^2).^2+(2 *
damp(m) * beta).^2);
```

```
    if m==1 plot(beta, D1,'k'), hold on, end
    if m==2 plot(beta, D1,'k:'),  end
    if m==3 plot(beta, D1,'k-.'),  end
    if m==4 plot(beta, D1,'k--'),  end
    if m==5 plot(beta, D1,'k','linewidth',2),  end
end
ylim([0 3])
plot(ones(1,1e3) * sqrt(2), linspace(0,3,1e3),'b:')
plot(sqrt(2),1,'bo','linewidth',2)
text(sqrt(2)+0.08,1.01,'\leftarrow(\surd2,1)')
xlabel('Frequency ratio \it\beta')
ylabel('\itX/Y')
legend('damping ratio = 0', 'damping ratio = 0.2', 'damping ratio =
0.4', ... 'damping ratio = 0.6', 'damping ratio = 0.8')
```

The following MATLAB program can be used to plot Fig. 3.16.

```
% For Fig. 3.16
clear all, close all
beta=linspace(0,3,1e3);
damp=[0 0.2 0.4 0.6 0.8];
for m=1:5
    D1=atan2(2 * damp(m) * (beta).^3,(1-beta.^2)) * 180/pi;
    if m==1 plot(beta, D1,'k'), hold on, end
    if m==2 plot(beta, D1,'k:'),  end
    if m==3 plot(beta, D1,'k-.'),  end
    if m==4 plot(beta, D1,'k--'),  end
    if m==5 plot(beta, D1,'k','linewidth',2),  end
end
ylim([-5 185])
xlabel('Frequency ratio \it\beta')
ylabel('angle (Deg)')
legend('damping ratio = 0', 'damping ratio = 0.2', 'damping ratio= 0.4', ...
'damping ratio = 0.6', 'damping ratio = 0.8',4)
```

Comment: From Fig. 3.15, it can be found that the motion transmitted is less than 1 for $\beta > \sqrt{2}$. Hence for vibration isolation, we must have $\frac{\omega}{\omega_n} > \sqrt{2}$, it means that the natural frequency ω_n must be as small as possible compared with excitation

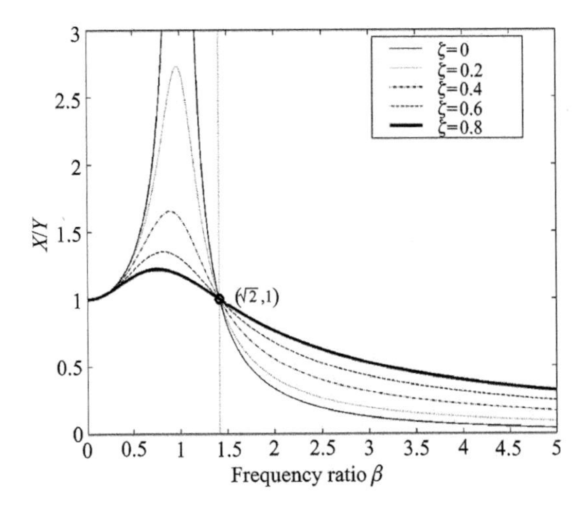

Fig. 3.15 The amplitude X/Y vary with the frequency ratio β under different damping ratio ζ

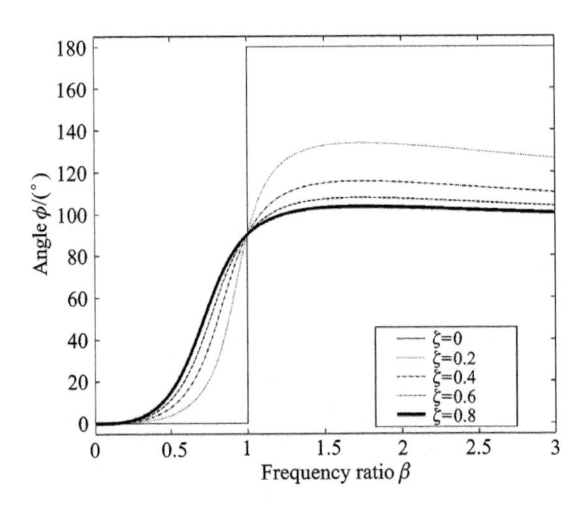

Fig. 3.16 The phase angle ϕ vary with the frequency ratio β under different damping ratio ζ

frequency ω.

The above solution can also be obtained if we assume a periodic input of

$$y = Y\sin(\omega t) \tag{3.69}$$

Substituting Eq. (3.69) into Eq. (3.62), the equation of motion can be rewritten as

$$m\ddot{x} + c\dot{x} + kx = kY\sin(\omega t) + c\omega Y\cos(\omega t) \tag{3.70}$$

Eq. (3.70) can be expressed as

$$m\ddot{x} + c\dot{x} + kx = A\sin(\omega t - \alpha) \tag{3.71}$$

where

$$A = Y\sqrt{k^2 + (c\omega)^2}, \quad \alpha = \arctan\frac{c\omega}{k} \tag{3.72}$$

Notice that Eq. (3.71) is in the same format now as the harmonic excitation case (see Section 3.1). The only differences are the amplitude of A instead of F_0 and an additional phase angle α. So we can find

$$x = X\sin(\omega t - \alpha - \phi) = \frac{Y\sqrt{k^2 + (c\omega)^2}}{\sqrt{(k - m\omega^2)^2 + (c\omega)^2}}\sin(\omega t - \alpha - \phi) \quad (3.73)$$

$$\alpha = \arctan\frac{c\omega}{k - m\omega^2} \quad (3.74)$$

From Eq. (3.73), the ratio of the amplitude can be obtained as follows:

$$\frac{X}{Y} = \frac{\sqrt{k^2 + (c\omega)^2}}{\sqrt{(k - m\omega^2)^2 + (c\omega)^2}} = \frac{\sqrt{1 + (2\zeta\beta)^2}}{\sqrt{(1-\beta^2)^2 + (2\zeta\beta)^2}} \quad (3.75)$$

and the phase angle ϕ can be expressed as

$$\phi = \arctan\frac{mc\omega^3}{k(k - m\omega^2) + (c\omega)^2} = \arctan\frac{2\zeta\beta^3}{1 - \beta^2 + (2\zeta\beta)^2} \quad (3.76)$$

Clearly, results shown in Eqs. (3.75) and (3.76) are the same solutions obtained using complex analysis (see Eqs. (3.67) and (3.68)), as expected.

Example 3.1

A spring-mounted body moves with velocity v along an undulating surface, as shown in Fig. 3.17. The body has a mass m and is connected to the wheel by a spring of stiffness k, and a viscous damper whose damping coefficient is c. The undulating surface has a wavelength L and an amplitude h.

Derive an expression for the ratio of amplitudes of the absolute vertical displacement of the body to the surface undulations.

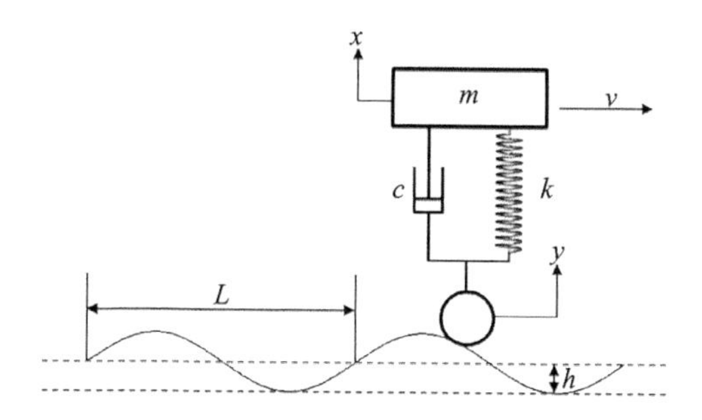

Fig. 3.17 A spring-mounted body moves along an undulating surface

Solution:

The system can be considered as free body diagram (see Fig. 3.18).

It is easy to find that

$$y = h\sin\frac{2\pi z}{L} \quad \text{and} \quad z = vt$$

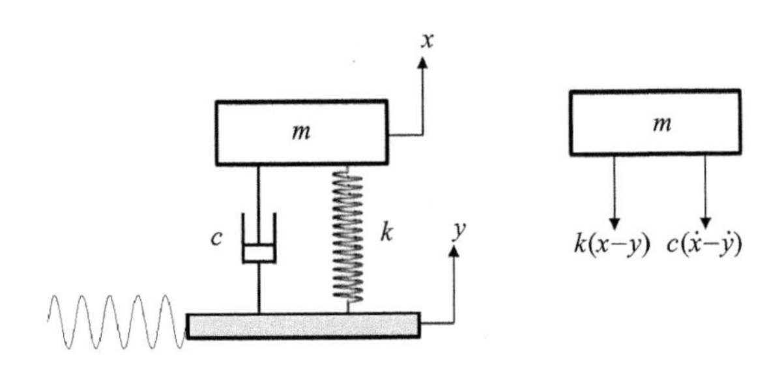

Fig. 3.18　The free body diagram

so

$$y = h\sin\left(\frac{2\pi v}{L}t\right) = h\sin(st) \quad \text{with} \quad s = \frac{2\pi v}{L}$$

By using Eq. (3.62), the equation of motion can be expressed as

$$m\ddot{x} + c\dot{x} + kx = ky + c\dot{y}$$

so we get

$$m\ddot{x} + c\dot{x} + kx = kh\sin(st) + csh\cos(st)$$

Above equation can be rewritten as

$$m\ddot{x} + c\dot{x} + kx = A\sin(st - \alpha)$$

where

$$A = h\sqrt{k^2 + (cs)^2} \quad \text{and} \quad \alpha = \arctan\frac{cs}{k}$$

Notice that the above equation is in the same format as Eq. (3.73). So the solution for this example can be expressed as

$$x = X\sin(st - \alpha - \phi) = \frac{h\sqrt{k^2 + (cs)^2}}{\sqrt{(k - ms^2)^2 + (cs)^2}}\sin(st - \alpha - \phi)$$

So the ratio of amplitudes of the absolute vertical displacement of the body to the surface undulations can be expressed as

$$\frac{X}{h} = \frac{\sqrt{k^2 + (cs)^2}}{\sqrt{(k - ms^2)^2 + (cs)^2}} \quad \text{with} \quad s = \frac{2\pi v}{L}$$

Example 3.2

Assume that there is an engine with mass 100 kg. The engine is installed by using anti-vibration mount which deflect 1 mm under the static weight of the engine. Assume that there is the damper of the mount is neglectable.

When running at 1 200 r/min, the amplitude of vibration of the set is measured to be 0.2 mm. If this vibration should be reduced to 0.1 mm without changing the mounting, what should we modify the system?

Solution:

First, the stiffness of the mount can be calculated as

$$k = \frac{mg}{\Delta_{st}} \quad \text{and} \quad \Delta_{st} = 1 \text{ mm}$$

so we can obtain $k = 980 \times 10^3$ N/m.

Moreover, for undamped mounts with zero initial conditions, from Eq. (3.67), it can be found that the amplitude of the engine:

$$X = \frac{Y}{|1-\beta^2|} = \frac{F_0}{|k(1-\beta^2)|} \quad \text{with} \quad \beta = \frac{\omega}{\omega_n}$$

Notice that $\omega_n^2 = \frac{k}{m}$, so the amplitude can be rewritten as $X = \frac{F_0}{|k - m\omega^2|}$.

Notice that the mounting cannot be changed, so the stiffness k should remain the same, and the only parameter is the mass of the engine can be modified.

Initially,

$$X_1 = \frac{F_0}{k - m_1\omega^2} = 0.2 \times 10^{-3} \text{ m}$$

and when changing the engine mass

$$X_2 = \frac{F_0}{k - m_2\omega^2} = 0.1 \times 10^{-3} \text{ m}$$

so

$$\frac{X_1}{X_2} = \left| \frac{k - m_2\omega^2}{k - m_1\omega^2} \right| = 2$$

and

$$\omega = \frac{2\pi \times 1\,200}{60} \text{ rad/s} = 40\pi \text{ rad/s} \quad \text{and} \quad m_1 = 100 \text{ kg}$$

Thus

$$\frac{X_1}{X_2} = \left| \frac{980 \times 10^3 - m_2 \times (40\pi)^2}{980 \times 10^3 - 100 \times (40\pi)^2} \right| = 2$$

It is easy to find that $m_2 = 137.9$ kg. It means that we can add mass of the engine to reduce the vibration.

3.6 Transmissibility of Vibration

Transmissibility functions are the transfer functions. They are particularly useful in the analysis of vibration isolation problems. Some delicate instruments must be isolated from their supports which may be subjected to vibrations, which is termed as motion isolation. On the other hand, vibration produced in machines should be isolated from the foundation so that adjoining structure is not set into vibration, which is termed as force isolation. So there are two types of transmissibilities, which will be discussed in this section.

3.6.1 Motion transmissibility

Consider a mechanical system that is supported through a suspension on a structure that may be subjected to undesirable motions (such as, guideway deflections, vehicle motions, seismic disturbances). Motion transmissibility determines the fraction of the support motion that is transmitted to the system through its suspension at different frequencies. It is defined as

$$\text{Motion transmissibility } T_{\mathrm{m}} = \frac{\text{System motion } X}{\text{Support motion } Y} \tag{3.77}$$

From Eq. (3.75), it can be find that the motion transmissibility T_{m} in Eq. (3.77) can be expressed as

$$T_{\mathrm{m}} = \frac{\sqrt{1+(2\zeta\beta)^2}}{\sqrt{(1-\beta^2)^2+(2\zeta\beta)^2}} \tag{3.78}$$

A schematic representation of the motion transmissibility mechanism is shown in Fig. 3.19. As discussed before (see Fig. 3.15 and Eq. (3.67)), we have known that the motion transmitted is less than 1 for $\beta > \sqrt{2}$. It means that the stiffness of spring should be satisfied

$$\sqrt{\frac{k}{m}} = \omega_{\mathrm{n}} < \frac{\sqrt{2}}{2}\omega \tag{3.79}$$

and

$$k < \frac{1}{2}\omega^2 m \quad \Rightarrow \quad T_{\mathrm{m}} < 1 \tag{3.80}$$

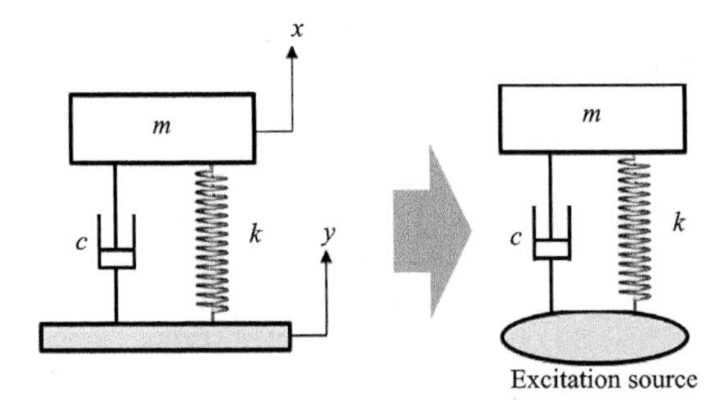

Fig. 3.19 The motion transmissibility mechanism

Example 3.3

Assume that the vibration on the floor in a building is simple harmonic motion at a frequency in the range 15—60 Hz. It is desired to install sensitive equipment in the building which must be insulated from floor vibration. The equipment is fastened to a small platform which is supported by a spring resting on the floor, each carrying an equal load. Only vertical motion is considered. The combined mass of the equipment and platform is 40 kg, and the equivalent viscous damping ratio of the suspension is 0.2.

Find the maximum value for the spring stiffness, if the amplitude of transmitted vibration is to be less than 10% of the floor vibration over the given frequency range.

Solution:

By using Eq. (3.78), the motion transmissibility T_m can be expressed as

$$T_m = \frac{\sqrt{1+(2\zeta\beta)^2}}{\sqrt{(1-\beta^2)^2+(2\zeta\beta)^2}}$$

$T_m < 0.1$ with $\zeta = 0.2$ is required, so

$$T_m = \frac{\sqrt{1+(2\times 0.2\times\beta)^2}}{\sqrt{(1-\beta^2)^2+(2\times 0.2\times\beta)^2}} = \frac{\sqrt{1+(0.4\beta)^2}}{\sqrt{(1-\beta^2)^2+(0.4\beta)^2}} < 0.1$$

From above equation, it is easy to find that

$$(1-\beta^2)^2+(0.4\beta)^2-100\left[1+(0.4\beta)^2\right] > 0$$

Thus $\beta > 4.72$ can be solved from above equation.

Notice that $\beta = \frac{\omega}{\omega_n}$ and $\omega \in [15\text{ Hz}, 60\text{ Hz}]$, so $\omega_n = \frac{\omega}{\beta} < \frac{15\times 2\pi}{4.72}$ for $T_m < 0.1$. It means that $\omega_n < 19.97$ rad/s should be satisfied.

And $k = m\omega_n^2$ with $m = 40$ kg, so $k < 15\ 935$ N/m is required.

Hence if stiffness of the spring is smaller than 15 935 N/m, the amplitude of the transmitted vibration will be less than 10% at frequencies above 15 Hz.

3.6.2 Force transmissibility

Consider a mechanical system that is supported on a rigid foundation through a suspension system, as shown in Fig. 3.20. If a forcing excitation $f(t) = F_0 \sin(\omega t)$ is applied to the system, the force is not directly transmitted to the foundation. The suspension system acts as a vibration isolation device. Force transmissibility determines the ratio of the forcing excitation that is transmitted to the support structure (foundation) through the suspension at different excitation frequencies. And it is defined as

$$\text{Force transmissibility } T_R = \frac{\text{Force transmitted to support } F_T}{\text{Applied force } F_0} \tag{3.81}$$

One important aspect of vibration analysis in application is determination of the steady-state amplitude of vibration under harmonic excitation. The other important factor is determination of the force $f_T(t)$ transmitted at steady state from the vibrating system to its supporting structure. We now address the later issue.

As shown in Fig. 3. 20, the force transmitted from the mass-spring-damper system to the ground consists of two components, that is, the damper and the spring forces.

$$f_T(t) = kx + c\dot{x} \tag{3.82}$$

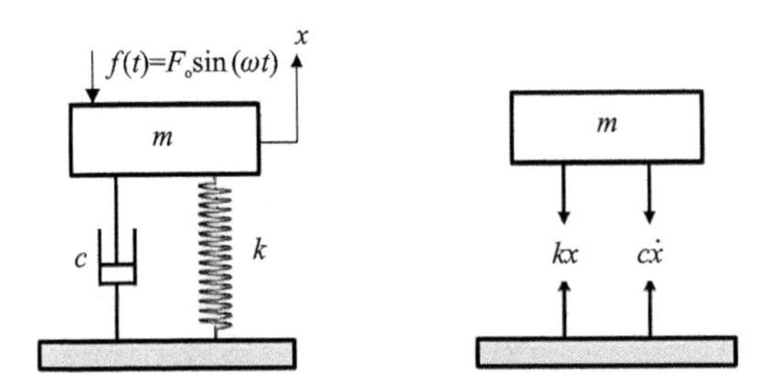

Fig. 3. 20 The force transmitted from the mass-spring-damper system

It has been shown in Section 3. 1 that the system at steady state vibrates with harmonic motion

$$x(t) = X\sin(\omega t - \phi) \tag{3.83}$$

Substituting Eq. (3. 83) into Eq. (3. 82), we have

$$f_T = c\omega X\sin\left(\omega t - \phi + \frac{\pi}{2}\right) + kX\sin(\omega t - \phi) \tag{3.84}$$

From Eq. (3. 84), it can be found that the force transmitted to the ground is also harmonic. Thus Eq. (3. 84) can be rewritten as

$$f_T = F_T\sin(\omega t - \alpha) \tag{3.85}$$

where F_T is constant, is the amplitude or maximum value of the force transmitted to the base.

Substituting Eq. (3. 83) into Eq. (3. 3) and by using Eq. (3. 85), it can be found that

$$\begin{aligned} & m\ddot{x} + c\dot{x} + kx \\ & = -m\omega^2 X\sin(\omega t - \phi) + c\omega X\sin\left(\omega t - \phi + \frac{\pi}{2}\right) + kX\sin(\omega t - \phi) \\ & = -m\omega^2 X\sin(\omega t - \phi) + F_T\sin(\omega t - \alpha) \\ & = F_0\sin(\omega t) \end{aligned} \tag{3.86}$$

Then Eq. (3. 86) has the graphical interpretation shown in Fig. 3. 21.

From Fig. 3. 21 and Eq. (3. 86), it is easy to find that

$$F_T = \sqrt{(c\omega X)^2 + (kX)^2} = X\sqrt{(c\omega)^2 + k^2} \tag{3.87}$$

and

$$\alpha = \arctan\frac{c\omega}{k - m\omega^2} - \arctan\frac{c\omega}{k} \tag{3.88}$$

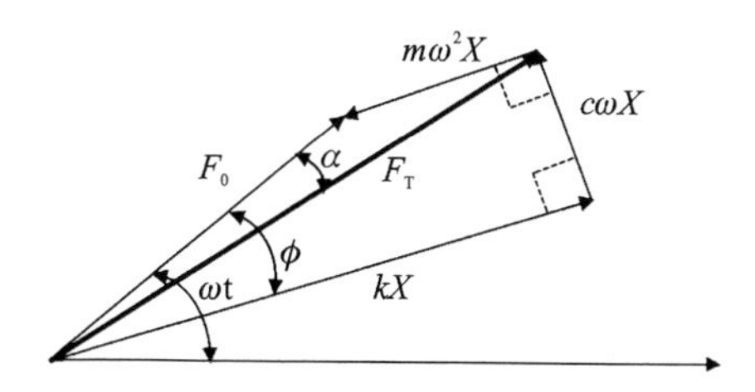

Fig. 3.21 The graphical interpretation for Eq. (3.86)

Notice that X in Eq. (3.87) has been given from Eq. (3.15), so F_T can be expressed as

$$F_T = \frac{F_0 \sqrt{(c\omega)^2 + k^2}}{\sqrt{(k - m\omega^2)^2 + (c\omega)^2}} \tag{3.89}$$

According to Eq. (3.81), the force transmissibility T_R can be written as

$$T_R = \frac{F_T}{F_0} = \frac{\sqrt{(c\omega)^2 + k^2}}{\sqrt{(k - m\omega^2)^2 + (c\omega)^2}} \tag{3.90}$$

Notice that $\beta = \frac{\omega}{\omega_n}$, $\zeta = \frac{c}{2\sqrt{km}}$ and $\omega_n^2 = \frac{k}{m}$, so Eq. (3.90) can be rewritten in dimensionless form:

$$T_R = \sqrt{\frac{1 + (2\zeta\beta)^2}{(1 - \beta^2)^2 + (2\zeta\beta)^2}} \tag{3.91}$$

Compare Eq. (3.91) with Eq. (3.78), it can be found that force transmissibility T_R is equal to motion transmissibility T_m.

3.7 Rotating Unbalance

Unbalance in rotating machines is a very common vibration excitation source. The rotating unbalance is a common problem in machinery. Consider a system of total mass M, with the rotating mass W, as shown in Fig. 3.22. The mass W rotates about point o with constant angular velocity ω and radius of rotation e. The system is supported by a spring of constant k and a damper of constant c. The common question is how to determine the amplitude of vibration of the mass M at steady state.

Without loss of generality, we may assume that the mass of the non-rotating structure is lumped at point O. Then we have position of $(M - W)$ is x, and position of W is $x + e\sin(\omega t)$. From Fig. 3.22, it can be found that the differential equation of motion can be expressed as

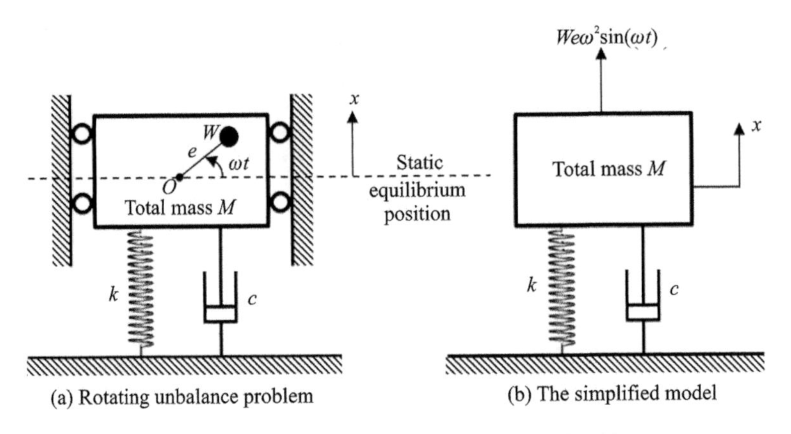

Fig. 3.22 Rotating unbalance problem and the simplified model

$$(M-W)\ddot{x}+W\frac{d^2[x+e\sin(\omega t)]}{dt^2}+c\dot{x}+kx=0 \tag{3.92}$$

Clearly, Eq. (3.92) can be rewritten as

$$M\ddot{x}-W\ddot{x}+W\ddot{x}+W\omega^2 e\sin(\omega t)+c\dot{x}+kx=0 \tag{3.93}$$

The differential equation of motion in Eq. (3.93) can be simplified in the form

$$M\ddot{x}+c\dot{x}+kx=W\omega^2 e\sin(\omega t) \tag{3.94}$$

The right hand side of the equation originates from the angular acceleration of the rotating unbalance in the x direction.

If we denote the particular solution $x_p(t)$ as follows:

$$x_p(t)=X\sin(\omega t-\phi) \tag{3.95}$$

Notice that $W\omega^2 e$ is constant. Hence substituting $W\omega^2 e$ for F_0 and M for m into Eq. (3.15), we can find that the steady-state amplitude of vibration is given by

$$X=\frac{W\omega^2 e}{\sqrt{(k-M\omega^2)^2+(c\omega)^2}} \tag{3.96}$$

Then the displacement amplitude ratio is

$$\frac{X}{X_0}=\frac{1}{\sqrt{(1-\beta^2)^2+(2\zeta\beta)^2}} \tag{3.97}$$

where $X_0=\frac{We\omega^2}{k}=\frac{We\beta^2}{M}$, the damping ratio $\zeta=\frac{c}{2\sqrt{km}}$, and $\beta=\frac{\omega}{\omega_n}$ is frequency ratio.

Eq. (3.96) can also be rewritten as

$$\frac{XM}{We}=\frac{\beta^2}{\sqrt{(1-\beta^2)^2+(2\zeta\beta)^2}} \tag{3.98}$$

and the phase angle ϕ in Eq. (3.95) can be expressed as

$$\phi = \arctan \frac{2\zeta\beta}{1-\beta^2} \tag{3.99}$$

Figs. 3.23 and 3.24 show the amplitude and phase of XM/We (in Eqs. (3.98) and (3.99)) under different damping ratios β. It can be found that the amplitude is low when the frequency ratio β is small.

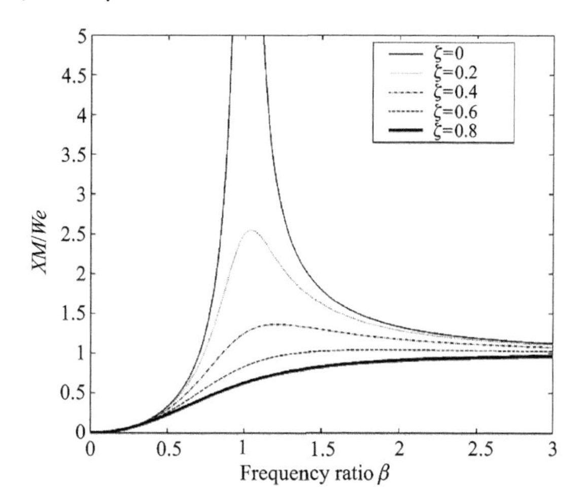

Fig. 3.23 The amplitude of XM/We in Eq. (3.98)

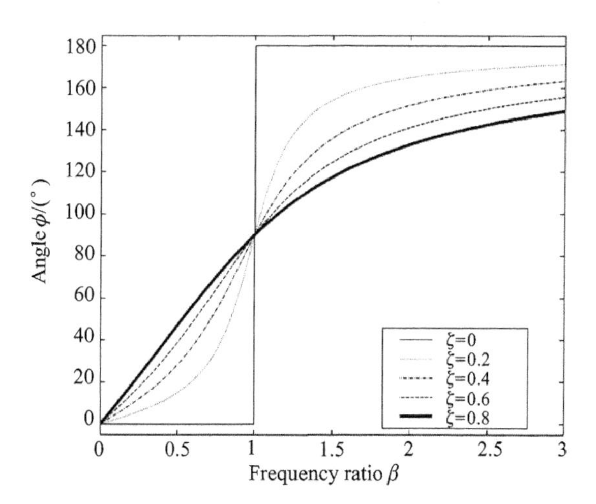

Fig. 3.24 The phase angle ϕ in Eq. (3.99)

The force transmitted because of the rotating imbalance can be expressed as

$$F = kx + c\dot{x} \tag{3.100}$$

The force transmissibility can be analyzed as before, such as

$$T_R = \frac{F}{F_0} = \frac{F}{We\omega^2} = \frac{F}{We(\beta\omega_n)^2} \tag{3.101}$$

thus

$$T_{\mathrm{R}}=\frac{F}{We\omega^2}=\frac{\beta^2\sqrt{1+(2\zeta\beta)^2}}{\sqrt{(1-\beta^2)^2+(2\zeta\beta)^2}} \tag{3.102}$$

The problem of reducing T_{R} is an interesting optimization problem. Fig. 3.25 shows the transmissibility T_{R} under different damping ratios. From Fig. 3.25, it can be found that, if T_{R} is less than one, then the system behaves like a vibration isolator, i. e., the ground receives less force than the rotating imbalance force.

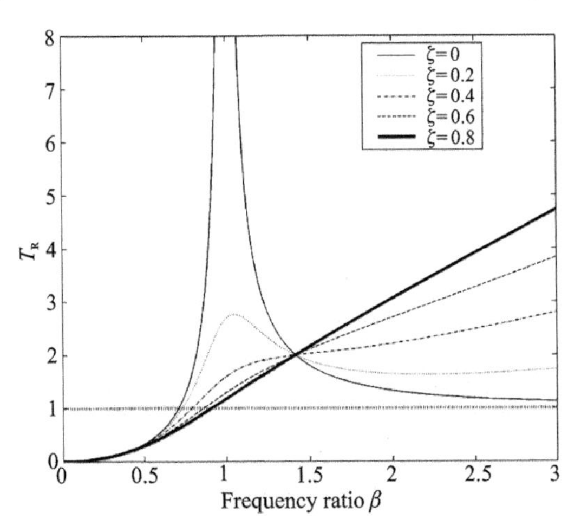

Fig. 3.25 T_{R} (due to rotating unbalance) under different damping ratios

Example 3.4

A top-loading washing machine executes the spin cycle at 240 r/min. The mass of the drum is 10 kg and it has a diameter of 0.6 m. The stiffness and damping coefficient of the mount are assumed as 379 N/m and 37.7 Ns/m respectively. If the mass of the clothes is 5 kg, what is the worst case possible for

(1) vibration amplitude?

(2) transmitted force?

Solution:

(1) According to Fig. 3.22 and Eq. (3.94), the equation of motion for rotating mass is

$$M\ddot{x}+c\dot{x}+kx=W\omega^2 e\sin(\omega t)$$

and the particular solution $x_{\mathrm{p}}(t)$ is

$$x_{\mathrm{p}}(t)=X\sin(\omega t-\phi)$$

Thus the steady-state amplitude of vibration is given by (see Eq. (3.96))

$$X=\frac{W\omega^2 e}{\sqrt{(k-M\omega^2)^2+(c\omega)^2}}$$

Notice that total mass $M=10\ \mathrm{kg}+5\ \mathrm{kg}=15\ \mathrm{kg}$ and the rotating mass $W=5\ \mathrm{kg}$.

The stiffness $k=379$ N/m and the damping coefficient $c=37.7$ Ns/m.
And $\omega=(2\pi\times 240/60)$ rad/s$=25.13$ rad/s.
For worst case, radius of rotation e = half of diamete r $=0.6$ m/2 = 0.3 m.
Thus the maximumamplitude of vibration is

$$X=\frac{W\omega^2 e}{\sqrt{(k-M\omega^2)^2+(c\omega)^2}}$$

$$=\frac{5\times 25.13^2\times 0.3}{\sqrt{(379-15\times 25.13^2)^2+(37.7\times 25.13)^2}}\ \mathrm{m}$$

$$=0.1\ \mathrm{m}$$

(2) The force transmitted because of the rotating imbalance is

$$F=kx+c\dot{x}$$

According to Eq. (3.102), we get

$$F=We\omega_n^2\frac{\beta^2\sqrt{1+(2\zeta\beta)^2}}{\sqrt{(1-\beta^2)^2+(2\zeta\beta)^2}}\quad \text{with}\quad \beta=\frac{\omega}{\omega_n}\quad \text{and}\quad \zeta=\frac{c}{2\sqrt{kM}}$$

Notice that $\omega_n=\sqrt{\frac{k}{M}}=\sqrt{\frac{379}{15}}$ rad/s$=5.03$ rad/s, so $\beta=5.0$.

The damping ratio $\zeta=\frac{c}{2\sqrt{kM}}=\frac{37.7}{2\sqrt{379\times 15}}=0.25$.

Hence $F=5\times 0.3\times 5.03^2\times\frac{5^2\times\sqrt{1+(2\times 0.25\times 5)^2}}{\sqrt{(1-5^2)^2+(2\times 0.25\times 5)^2}}$ N$=105.87$ N.

3.8 MATLAB Examples for Forced SDOF System

3.8.1 Harmonic excitation of undamped SDOF systems

For harmonic excitation cases, the responses can be solved by using Eqs. (3.32) and (3.35) under different situations. The following MATLAB program can be used to solve all harmonic excitation cases.

MATLAB Codes	Comments
% Force vibration for undamped SDOF system	
clear all, close all	
m=1;	The mass
f0=1;	The excitation force
wn=10;	Natural frequency of the SDOF system
wdr=10;	The excitation frequency

(continued)

x0=0;	Initial displacement
v0=0;	Initial velocity
k=wn^2 * m;	The stiffness
beta=wdr/wn;	The frequency ratio
t=linspace(0,10,1e3);	
xh=x0 * cos(wn * t)+v0/wn * sin(wn * t);	The free response
if wn ~= wdr	
xp=f0/k/(1-beta^2) * (sin(wdr * t)-beta * sin(wn * t));	The force response for excitation frequency is NOT equal to natural frequency
else	
xp=-wn * f0/2/k * t. * cos(wn * t);	The force response for excitation frequency is equal to natural frequency
end	
xt=xh+xp;	The total response
figure(1),plot(t,xt,'k')	Plot the results
ylabel('Response x')	
xlabel('Time(second)')	

By using above program, Fig. 3.26 shows the resonance response when $\omega = \omega_n = 10$ rad/s. Fig. 3.27 shows the beating response when $\omega = 11$ rad/s and $\omega_n = 10$ rad/s. And Fig. 3.28 shows the effect of varying excitation frequency ω for a given natural frequency. In these calculations, $v_0 = 0$; $x_0 = 0$; $F_0 = 1$ and $m = 1$ are used. Making the initial conditions zero allows us to better see the effects of

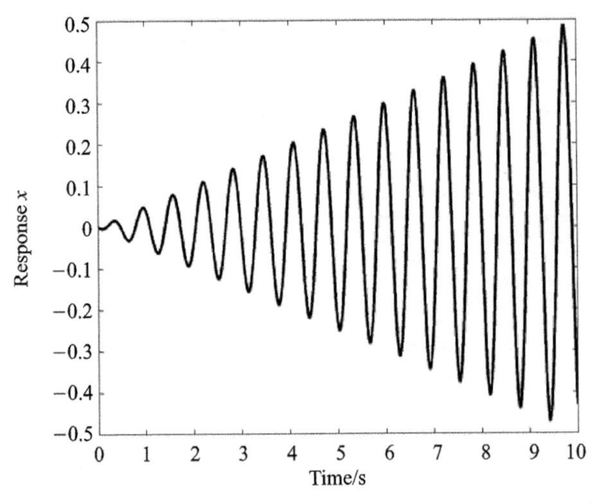

Fig. 3.26　The resonance response when $\omega = \omega_n = 10$ rad/s

varying frequencies. A simple (and natural) modification to above program would be to obtain the different results. From a practical point of view, though, introducing too many plots to the figures would make the plot unreadable, so care must be taken.

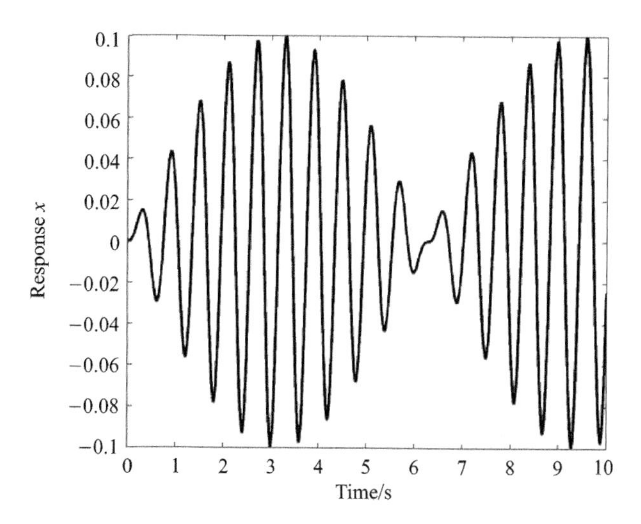

Fig. 3.27 The beating response when $\omega = 11$ rad/s and $\omega_n = 10$ rad/s

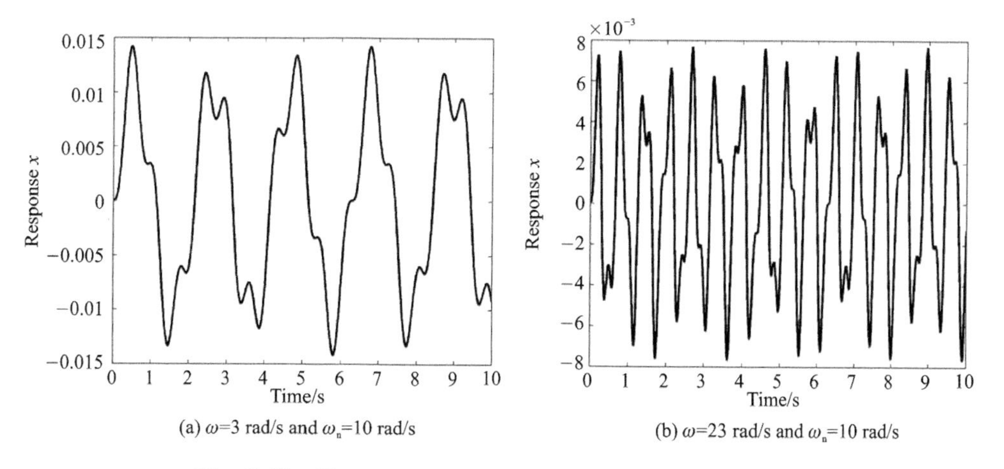

(a) ω=3 rad/s and ω_n=10 rad/s (b) ω=23 rad/s and ω_n=10 rad/s

Fig. 3.28 The response under different excitation frequencies

3.8.2 Rotating unbalance vibration of SDOF systems

For the purpose of creating a program in MATLAB, the initial conditions were assumed to be zero. The following MATLAB program can be used to calculate the response of the unbalance vibration system.

```
%This program solves for the response of a single-degree-of-freedom
%system having a rotating unbalance.
%The equations of motion are valid only for zero initial conditions.
mo=3;
m=7;
e=0.1;
wr=4;
zeta=0.05;
tf=10;
t=0:tf/1000:tf;
wn=2;
wd=wn * sqrt(1-zeta^2);
r=wr/wn;
X=mo * e/m * (r^2/sqrt((1-r^2)^2+(2 * zeta * r)^2));
phi=atan2(2 * zeta * r,(1-r^2));
Z1=(-zeta * wn+wr * cot(phi))/wd;
Z2=sqrt((zeta * wn)^2-2 * zeta * wn * wr * cot(phi)+(wr * cot(phi))^2+
wd^2)/wd;
Z=Z1+Z2;
theta=2 * atan(Z);
Anum=X * (wd * sin(phi)-Z * zeta * wn * sin(phi)+Z * wr * cos(phi));
Aden=Z * wd;
A=Anum/Aden;
xh=A * exp(-zeta * wn * t). * sin(wd * t+theta);
xp=X * sin(wr * t-phi);
x=xp+xh;
figure(1)
plot(t,x(:))
title(['Rotating unbalance with wr=',num2str(wr),'    wn=' num2str(wn),'
and \zeta=',num2str(zeta)]);
xlabel('Time(second)')
```

Fig. 3. 29 shows the time response by using above program with different excitation frequencies. In this case, $M = 7$, $W = 3$, and $e = 0.1$ are used.

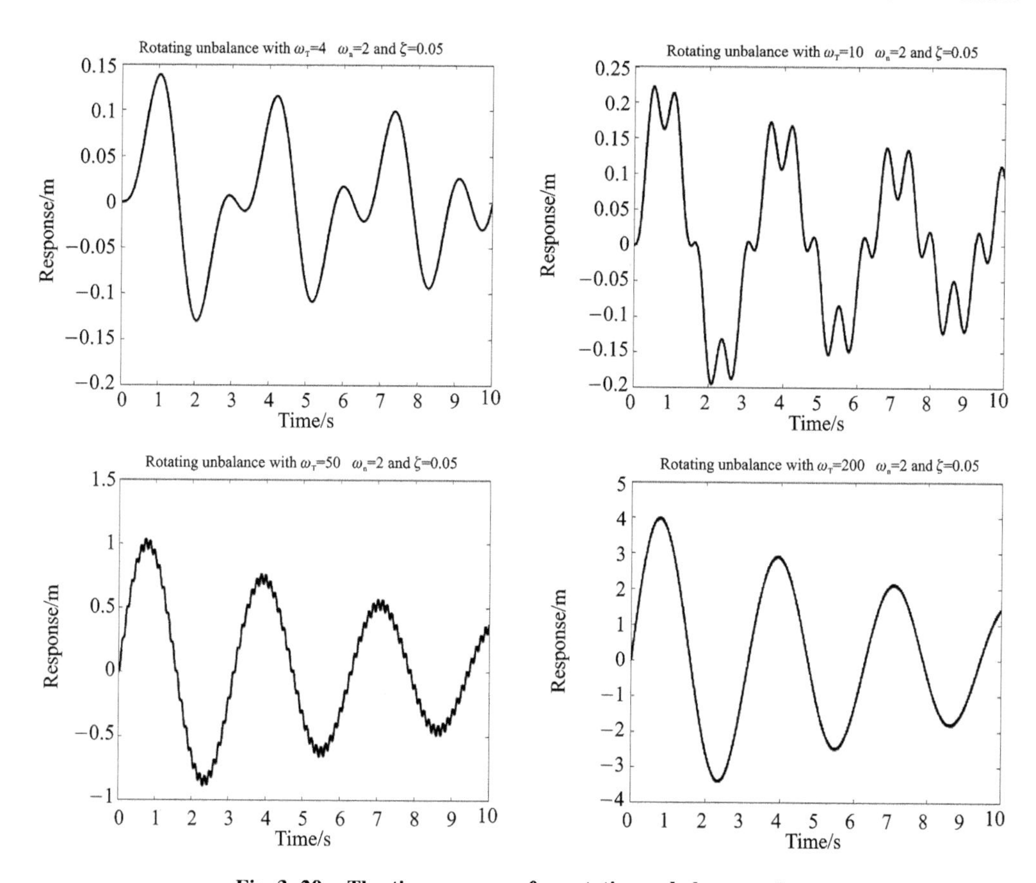

Fig. 3. 29 The time response for rotating unbalance system

Questions

3. 1 To understand the nature of the forced response for the undamped systems, plot the solution of a forced response of Eq. (3. 1) with $\omega = 2$ rad/s, given by Eq. (3. 29) for a variety of values of the initial conditions and ω_n as given in the following table.

Case	x_0	v_0	F_0	ω_n
1	0. 1	0. 1	0. 1	1
2	−0. 1	0. 1	0. 1	1
3	0. 1	0. 1	1. 0	1
4	0. 1	0. 1	0. 1	2. 1
5	1	0. 1	0. 1	1

3.2 Compute the total response of a undamped spring-mass system with the following values: k = 1 000 N/m, m = 10 kg, subject to a harmonic force of magnitude F_0 = 1 N and frequency of 2 Hz, and initial conditions given by x_0 = 0.01 m and v_0 = 0.01 m/s.

3.3 Consider the system in Fig. 3.30 with k = 5 000 Nm, m = 100 kg under the excitation force $f(t)=10\sin 2t$. Write the equation of motion and calculate the response assuming that

(1) the system is initially at rest;

(2) the system has an initial displacement of 0.01 m;

(3) the system has an initial displacement of 0.01 m and initial velocity of 0.1 m/s.

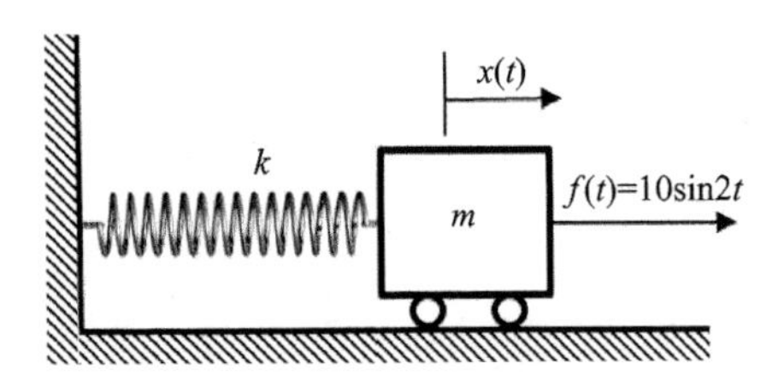

Fig. 3.30 A SDOF system

3.4 Consider the system in Fig. 3.31, write the equation of motion and calculate the response assuming that the system is initially at rest for the values $k_1 = k_2$ = 100 N/m, k_3 = 500 N/m and m = 10 kg.

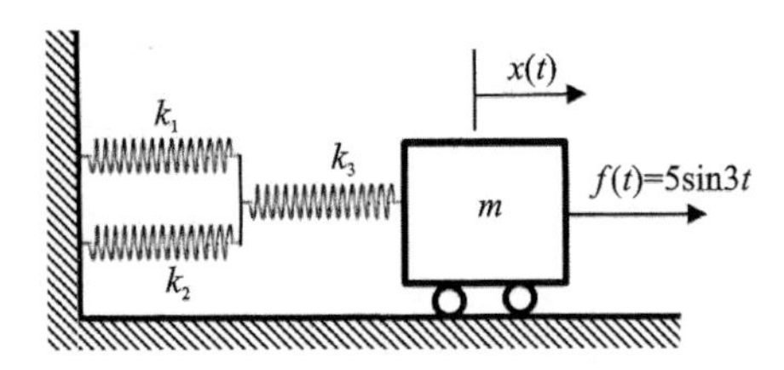

Fig. 3.31 A SDOF system with series springs

3.5 Consider the system in Fig. 3.32, write the equation of motion and calculate the response assuming that the system is initially at rest for the values θ = 35°, k = 100 N/m and m = 5 kg.

3.6 Can the response of the follow equation

$$m\ddot{x} + kx = F_0\cos(\omega t)$$

oscillates at only one frequency (ω)? If can, please compute the initial conditions for this case.

3.7 An airfoil is mounted in a wind tunnel for the purpose of studying the aerodynamic properties of the airfoil's shape. The airfoil is assumed as a rigid body with the rotational inertia J, as shown in Fig. 3.33. If the magnitude of the

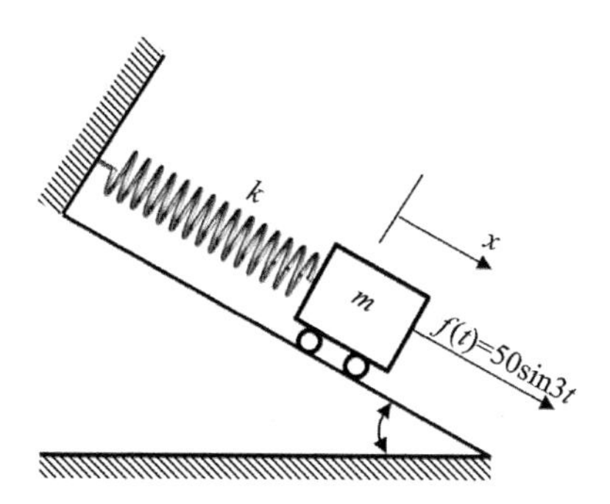

Fig. 3.32 A spring-mass system on frictionless surface

angular deflection should be less than 5°, compute the stiffness of the rotational spring. Assume that the initial conditions are zero.

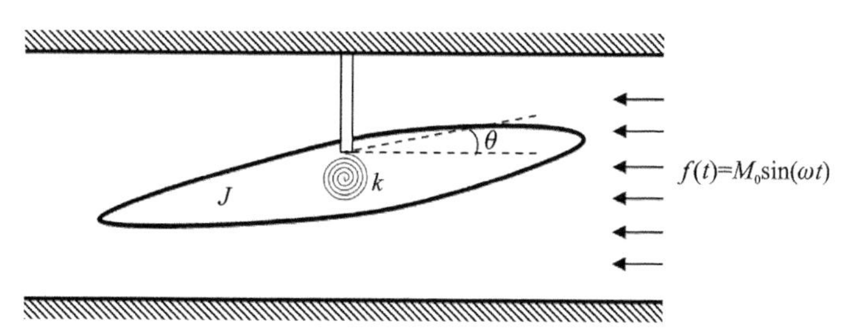

Fig. 3.33 Vibration model of a wing in a wind tunnel

3.8 According to Eq. (3.38), if the excitation force $f(t) = F_0\cos(\omega t)$, the response of the underdamped system can be expressed as

$$x(t) = A\exp(-\zeta\omega_n t)\sin(\omega_d t + \phi) + X\cos(\omega t - \phi)$$

(1) Calculate the constants A and ϕ for arbitrary initial conditions x_0 and v_0;

(2) Compare the solution in (1) with the transient response obtained in case of no forcing function (such as $F_0 = 0$).

3.9 Plot the solution of Question 3.8 for the case that $m = 1$ kg, $\zeta = 0.01$, $\omega_n = 1$ rad/s, $F_0 = 1$ N and $\omega = 10$ rad/s, with initial conditions $x_0 = 0.1$ m and $v_0 = 0.1$ m/s.

3.10 A 100 kg mass is suspended by a spring of stiffness 3 000 N/m with a viscous damping constant of 500 Ns/m. The mass is initially at rest and in equilibrium. Calculate the steady-state displacement amplitude and phase if the mass is excited by a harmonic force of 10 N at 3 Hz and 10 Hz, respectively.

3.11 Compute the forced response of a spring-mass-damper system with the following values: $c = 100$ kg/s, $k = 1\,000$ N/m, $m = 100$ kg, subject to a harmonic force of magnitude $F_0 = 25$ N and frequency of 10 rad/s and initial conditions of $x_0 = 0.01$ m and $v_0 = 0.01$ m/s. Plot the response. How long does it take for the transient part to die off?

3.12 A spring-mass-damper system is shown in Fig. 3.34, with $k = 1\,000$ Nm, $m = 10$ kg under the excitation force $f(t) = 25\cos 18t$. Compute a value of the damping coefficient c such that the steady-state response amplitude of the system is 0.01 m.

3.13 Compute the response of the system in Fig. 3.35, if the system is initially at rest for the values $k_1 = k_2 = k_3 = 500$ N/m, $c = 20$ kg/s and $m = 100$ kg.

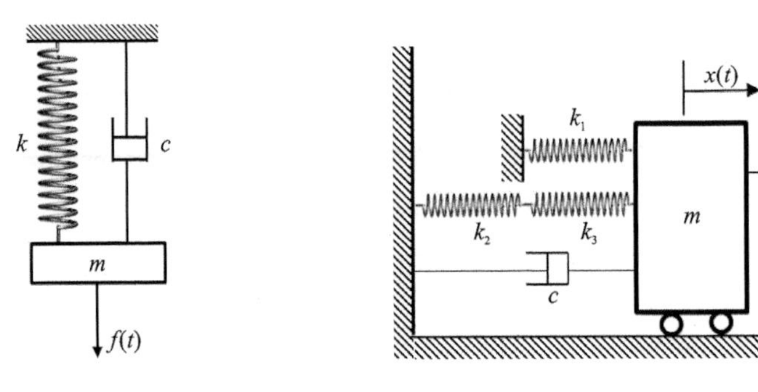

Fig. 3.34 A spring-mass-damper system **Fig. 3.35 Damped vibration system**

3.14 Consider the system in Fig. 3.36. Assume that $k = 2\,000$ N/m, $c = 25$ kg/s, $m = 25$ kg and $F(t) = 50\sin 8t$ N. Compute the steady-state response assuming the system starts from rest. Also use the small angle approximation.

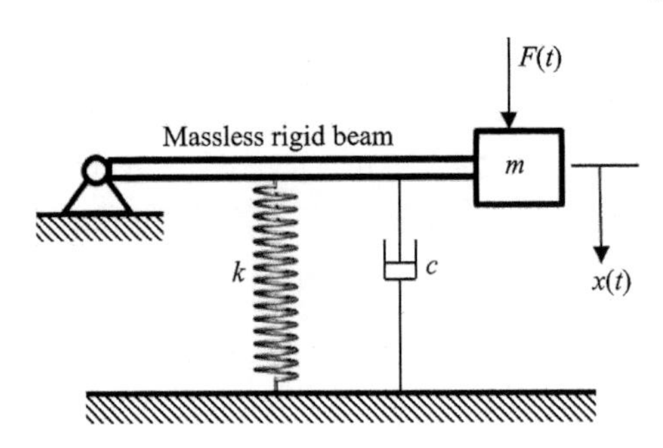

Fig. 3.36 A massless beam with spring and damper

3.15 Recall the system of in Question 3.7, repeated here as Fig. 3.37 with the effects of damping indicated. The physical constants are $J = 10$ kg · m^2, $k = 1\,000$ N/m, and the applied moment M_0 is 10 Nm at 1 Hz acting through the distance $l = 0.5$ m.

(1) Compute the magnitude of the steady-state response if the measured damping ratio of the spring system is $\zeta = 0.01$.

(2) Compare this to the response for the case where the damping is not modeled ($\zeta = 0$).

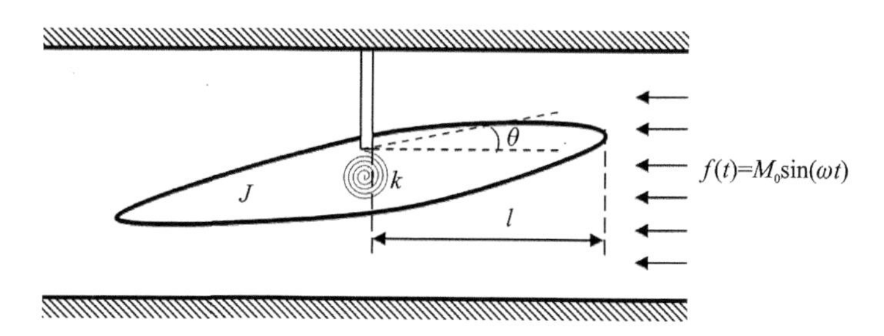

Fig. 3.37 Model of an airfoil in at wind tunnel including the effects of damping

3.16 A body of mass 10 kg is suspended by a spring of stiffness of 1 000 N/m and dashpot of damping constant 1 000 Ns/m. Vibration is excited by a harmonic force of amplitude 10 N and a frequency of 10 Hz. Calculate

(1) the amplitude of the displacement for the mass;

(2) the phase angle between the displacement and the excitation force.

3.17 A device with weight of 1 000 N rests on a support, as shown in Fig. 3.38. The spring deflects about 10 mm as a result of the weight of the device. The base under the support is flexible and harmonically vibration with an amplitude of 1 mm. Calculate the transmitted force and the amplitude of the transmitted displacement:

(1) If the damping ratio of system is $\zeta = 0.01$;

(2) If the damping ratio of system is $\zeta = 0.1$.

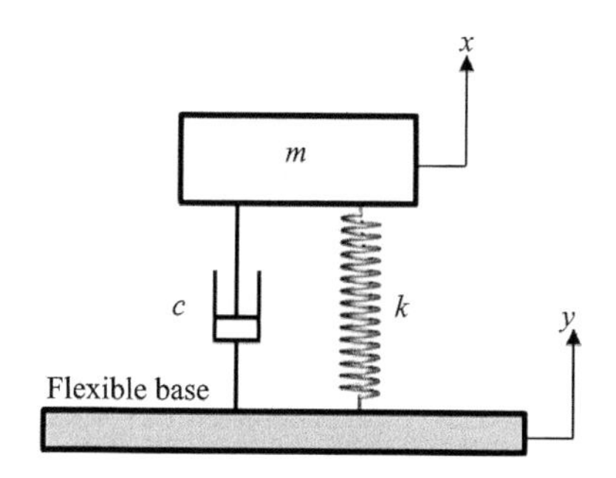

Fig. 3.38 A mass on flexible base

3.18 From Fig. 3.15, it can be found that the point $(\sqrt{2}, 1)$ corresponds to the value of the dynamic magnification factor $D = 1$. Please verify this conclusion.

3.19 Consider the base excitation problem for the configuration shown in Fig. 3.39. In this case, the base motion is a displacement transmitted through a pure damper. Derive an expression for the force transmitted to the support in steady state.

3.20 A very common example of base motion is the SDOF model of an automobile driving over a rough road. The road is modeled as providing a base

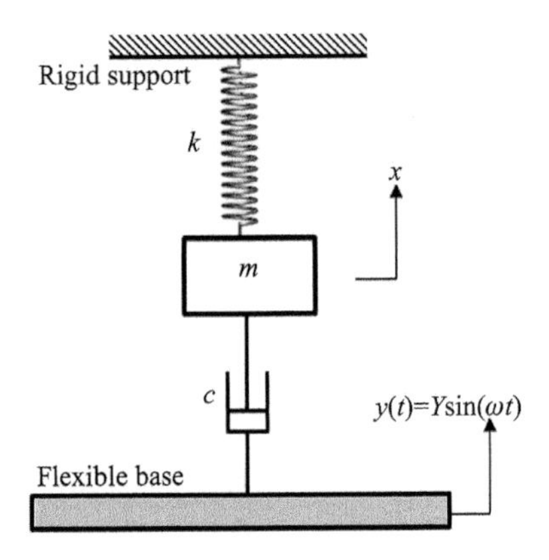

Fig. 3.39 A base excitation problem

motion displacement of $y(t)=0.01\sin 8t$ m. The suspension provides an equivalent stiffness of $k = 10^5$ N/m, a damping coefficient of $c = 40\times10^3$ kg/s and a mass of 1 000 kg. Determine the amplitude of the absolute displacement of the automobile body.

3.21 Assume that a vibrating mass of 100 kg is mounted on a massless support by a spring of stiffness 10×10^3 N/m and a damper. It is found that the amplitude of the mass is 10 mm, when the support vibration has maximum amplitude of 2 mm (at resonance). Calculate the damping constant and the amplitude of the force on the base.

3.22 A lathe can be modeled as an electric motor mounted on a steel table. Assume that the total mass (table plus the motor) is 100 kg. The mass of the rotating parts is 5 kg at the distance 0.1 m from the center. The damping ratio of the system is assumed as $\zeta=0.06$ (viscous damping) and its natural frequency is 10 Hz. Calculate the amplitude of the steady-state displacement of the motor, assuming rotation frequency $\omega = 25$ Hz.

3.23 Consider a system with rotating unbalance as illustrated in Fig. 3.21. Suppose the deflection at 1 750 r/min is measured to be 0.05 m and the damping ratio is measured to be $\zeta = 0.1$. The out-of-balance mass is 10% of the total mass. Locate the unbalanced mass by computing e.

3.24 When a helicopter blade is rotating at 360 r/min, it is found that the fuselage vibrates strongly and resonance occurs at this rotating speed. Assume that there is a device on the fuselage, its motion transmissibility T_d should be less than 0.2, try to find the static deformation of the isolating spring under the self-weight of the device.

3.25 A mass-spring-damper system is shown in Fig. 3.40 under excitation force $f_0 = 15$ N, $\omega = 10$ rad/s. Assume that there are four configurations:

(1) $m = 15$ kg, $k = 400$ N/m and $c = 0$;

(2) $m = 22.5$ kg, $k = 400$ N/m and $c = 0$;

(3) $m = 15$ kg, $k = 500$ N/m and $c = 0$;

(4) $m = 15$ kg, $k = 400$ N/m and $c = 180$ Ns/m.

Which is the best choice to minimize the force transmissibility?

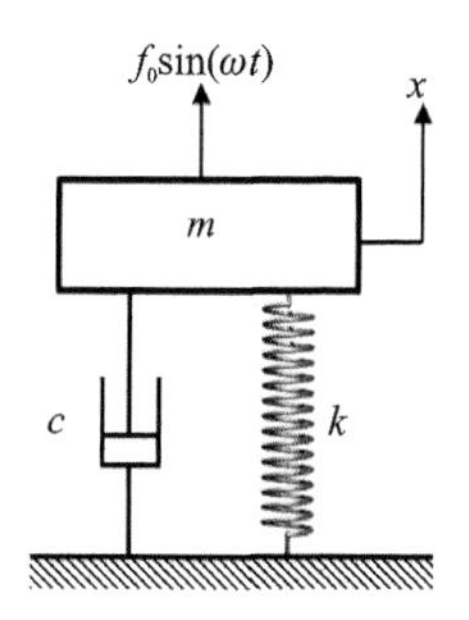

Fig. 3.40 A mass-spring-damper system

3.26 One of the equations below describes a undamped mass-spring system undergoing resonance. Identify the equation, and find its general solution.

(1) $\ddot{x} + 16x = 5\cos 8t$;

(2) $\ddot{x} + 2\dot{x} + 4x = 2\sin 2t$;

(3) $\ddot{x} + 16x = 7\cos 2t$.

3.27 Find the value(s) of k, such that the undamped mass-spring system described by each of the equations below is undergoing resonance.

(1) $8\ddot{x} + kx = 5\sin 6t$;

(2) $3\ddot{x} + kx = -\pi\cos t$.

Chapter 4 Vibration of SDOF Systems under General Excitation

Some structures are subjected to shock or impulse loads arising from suddenly applied, non-periodic, short-duration exciting forces or periodic forces. In this chapter, the response of a SDOF system under general excitation will be presented.

4.1 The Impulse Response

The impulse, sometimes termed as delta function, is a fictitious function, formally defined in the following way

$$\delta(t-t_0)=\begin{cases}\infty & t=t_0 \\ 0 & t\neq t_0\end{cases} \tag{4.1}$$

From Eq. (4.1), it can be found that this function has singularity at $t=t_0$, as shown in Fig. 4.1.

Notice that Eq. (4.1) does not really lead to any legitimate function. However, the delta function can be considered as the box function of unit area when the limit value ε tends to 0, as shown in Fig. 4.2.

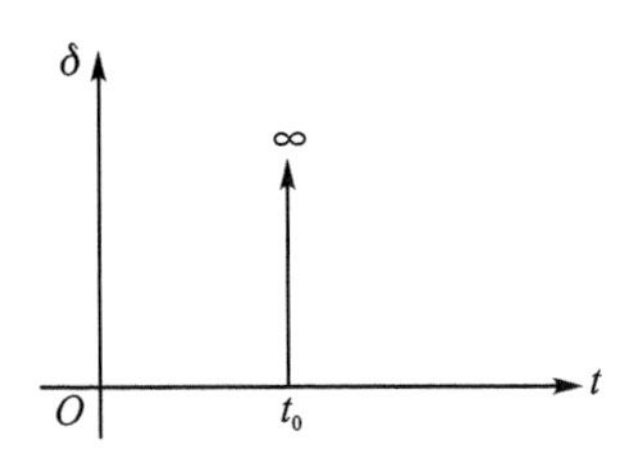

Fig. 4.1 The delta function

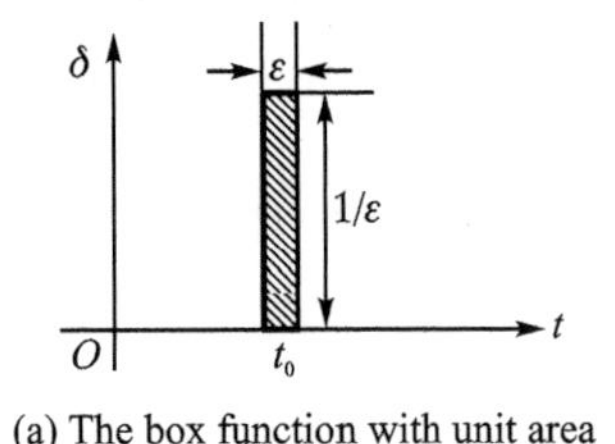

(a) The box function with unit area

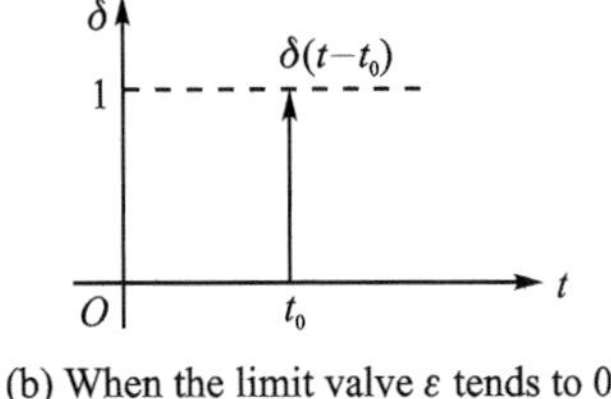

(b) When the limit valve ε tends to 0

Fig. 4.2 The delta function with a limit value

So a more accurate definition of the delta function can be obtained as follows:

$$\int_{-\infty}^{+\infty}\delta(t-t_0)\mathrm{d}t=1 \tag{4.2}$$

Assume that a mass-spring-damper system is applied with a unit impact at the time $t=t_0$, as shown in Fig. 4.3. This impact may be expressed mathematically by $\delta(t-t_0)$. The question is how to determine the impulse response $h(t-t_0)$ due to the impact $\delta(t-t_0)$. Assume that the system is initially at rest (initial conditions $x(0)=v(0)=0$). Clearly, the motion equation can be expressed as

$$m\ddot{x}+c\dot{x}+kx=\delta(t-t_0) \tag{4.3a}$$

and the initial conditions can be written as

$$x(t)=\dot{x}(t)=0 \quad \text{for} \quad t<t_0 \tag{4.3b}$$

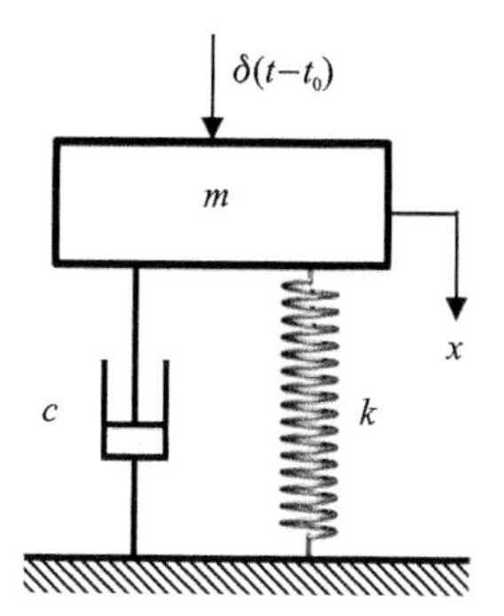

Fig. 4.3 A mass-spring-damper system with unit impact force

Clearly, Eq. (4.3b) implies that the system has no motion for $t<t_0$. Thus only the dynamic of the system for $t>t_0$ is waiting to determine.

Notice that **there is no applied force after the impact**, the response of the system after the impact can be regarded as a free vibration system, which is governed by

$$m\ddot{x}+c\dot{x}+kx=0, \quad t>t_0 \tag{4.4}$$

$$x(t_0^+)=A \quad \text{and} \quad \dot{x}(t_0^+)=B \tag{4.5}$$

where t_0^+ represents the time immediately after the impact, and A and B are unknown constants representing the displacement and velocity of the mass at the time $t=t_0^+$, respectively. Similarly, the time immediately before the impact can be denoted as $t=t_0^-$.

From Eqs. (4.4) and (4.5), to obtain the response of the system shown in Fig. 4.3, the unknown constants A and B should be determined. According to Eq. (4.2), we get

$$\int_{-\infty}^{+\infty}\delta(t-t_0)\mathrm{d}t=\int_{t_0^-}^{t_0^+}\delta(t-t_0)\mathrm{d}t=1 \tag{4.6}$$

Notice that the spring and damper forces are non-impulsive forces, the following equation can be found by using the Newton's second law during the impact:

$$\delta(t-t_0)=m\ddot{x} \tag{4.7}$$

Substituting Eq. (4.7) into Eq. (4.6), we get

$$\int_{t_0^-}^{t_0^+}m\ddot{x}\,\mathrm{d}t=1 \tag{4.8}$$

Integrating Eq. (4.8) yields

$$\int_{t_0^-}^{t_0^+}m\ddot{x}\,\mathrm{d}t=m\dot{x}\,\Big|_{t=t_0^-}^{t_0^+}=m\dot{x}(t_0^+)-m\dot{x}(t_0^-)=1 \tag{4.9}$$

Notice that the system is at rest before the impact, that is, $\dot{x}(t_0^-)=0$. Hence Eq. (4.9) gives

$$m\dot{x}(t_0^+)=1 \quad \text{and} \quad \dot{x}(t_0^+)=\frac{1}{m}=B \tag{4.10}$$

This important result shows that the velocity gained by a unit impact equals the inverse of the mass (independent of the spring constant k and the damper constant c). In fact, this is the fundamental principle of impulse and momentum: **"The change in momentum equals the impulse of the acting force".**

Integrating Eq. (4.10), we get

$$\int_{t_0^-}^{t_0^+} m\dot{x}\,\mathrm{d}t=\int_{t_0^-}^{t_0^+} 1\mathrm{d}t=0 \tag{4.11}$$

From Eq. (4.11), we have

$$mx\Big|_{t=t_0^-}^{t_0^+}=t\Big|_{t=t_0^-}^{t_0^+}=0 \tag{4.12}$$

Eq. (4.12) can be rewritten as

$$mx(t_0^+)-mx(t_0^-)=t_0^+-t_0^-=0 \tag{4.13}$$

Notice that $x(t_0^-)=0$ according to Eq. (4.3b). From Eq. (4.13), it can be found that

$$mx(t_0^+)=0 \quad \text{and} \quad x(t_0^+)=0=A \tag{4.14}$$

From Eq. (4.14), it can be found that the mass cannot gain finite displacement by an impulsive force.

In summary, with Eqs. (4.4), (4.5), (4.10) and (4.14), the impulse response of the mass-spring-damper system is governed by the following equations:

$$m\ddot{x}+c\dot{x}+kx=0, \quad t>t_0 \tag{4.15}$$

$$x(t_0^+)=0 \quad \text{and} \quad \dot{x}(t_0^+)=\frac{1}{m} \tag{4.16}$$

Normally, we would like to set the initial conditions at $t=0$. We can apply to Eqs. (4.15) and (4.16) by using $(t-t_0)$ as variable, such as

$$m\ddot{x}(t-t_0)+c\dot{x}(t-t_0)+kx(t-t_0)=0, \quad t-t_0>0 \tag{4.17}$$

$$x(0)=0 \quad \text{and} \quad \dot{x}(0)=\frac{1}{m} \tag{4.18}$$

Assume that the damping ratio $\zeta<1$, the solution can be obtained by using Eq. (2.123) in Chapter 3.

$$x(t-t_0)=\begin{cases}0 & t-t_0<0\\ \dfrac{\exp[-\zeta\omega_n(t-t_0)]}{m\omega_d}\sin[\omega_d(t-t_0)] & t-t_0\geqslant 0\end{cases} \tag{4.19}$$

Eq. (4.19) determines the response of an underdamped system ($\zeta<1$) due to an impulse force at the time $t=t_0$.

In vibration analysis, we always use $h(t)$ to replace $x(t)$ in Eq. (4.19), and $h(t)$ is termed as unit impulse response function.

$$h(t-t_0)=\begin{cases}0 & t-t_0<0\\ \dfrac{\exp[-\zeta\omega_n(t-t_0)]}{m\omega_d}\sin[\omega_d(t-t_0)] & t-t_0\geqslant 0\end{cases} \tag{4.20}$$

Clearly, similar expressions can be simply obtained from Eq. (2.123) for the cases where $\zeta \geqslant 1$ or $\zeta = 0$.

4.2 The Principle of Superposition

A system is called a linear system if its response to the excitation $f(t) = f_1(t) + f_2(t)$ can be obtained by superimposing the response due to $f_1(t)$ and that corresponding to $f_2(t)$. Consider now the mass-spring-damper system with two excitations, $f_1(t)$ and $f_2(t)$, shown in Fig. 4.4.

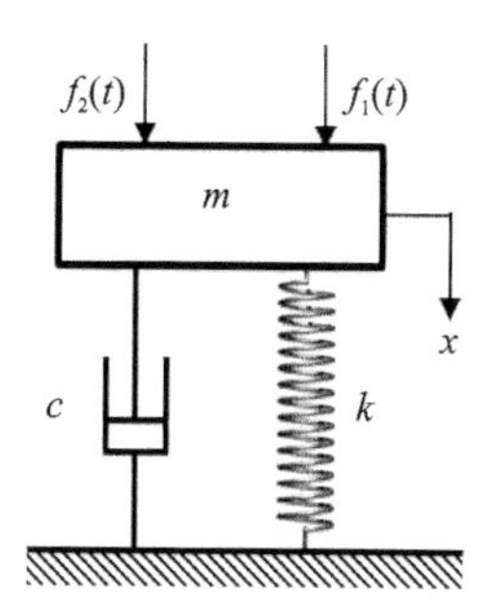

Fig. 4.4 A linear system with the excitations $f_1(t)$ and $f_2(t)$

The differential equation of motion is

$$m\ddot{x}+c\dot{x}+kx=f_1(t)+f_2(t) \tag{4.21}$$

Assume that $x_1(t)$ is the solution of

$$m\ddot{x}_1+c\dot{x}_1+kx_1=f_1(t) \tag{4.22}$$

and $x_2(t)$ is the solution of

$$m\ddot{x}_2+c\dot{x}_2+kx_2=f_2(t) \tag{4.23}$$

Then

$$x(t)=x_1(t)+x_2(t) \tag{4.24}$$

is the solution of Eq. (4.21).

Proof:

Adding Eq. (4.22) to Eq. (4.23), we get

$$(m\ddot{x}_1+c\dot{x}_1+kx_1)+(m\ddot{x}_2+c\dot{x}_2+kx_2)=f_1(t)+f_2(t) \tag{4.25}$$

Rearranging Eq. (4.25), we have

$$m(\ddot{x}_1+\ddot{x}_2)+c(\dot{x}_1+\dot{x}_2)+k(x_1+x_2)=f_1(t)+f_2(t) \tag{4.26}$$

According to Eq. (4.24), Eq. (4.26) can be rewritten as

$$m\ddot{x} + c\dot{x} + kx = f_1(t) + f_2(t) \tag{4.27}$$

which is the same as Eq. (4.21). According to above proof, it is easy to found that the mass-spring-damper system, discussed in Chapters 2 and 3, is a linear system. It should be noticed that most real life systems are nonlinear. To simplify the analysis we usually make **"linearisation"**.

Example 4.1

Consider the simple pendulum shownin Fig. 4.5, what is motion equation for the simple pendulum under small and large angles?

Solution:

Applying the Newton's second law, the differential equation of motion for the simple pendulum can be obtained from

$$mL\ddot{\theta} = -mg\sin\theta \quad \text{or} \quad L\ddot{\theta} + g\sin\theta = 0 \tag{4.28}$$

Fig. 4.5 A simple pendulum

This is a nonlinear system in the sense that the superposition principle does not hold with respect to Eq. (4.28). Assume that θ_1 and θ_2 are solutions of

$$L\ddot{\theta}_1 + g\sin\theta_1 = 0 \tag{4.29}$$

and

$$L\ddot{\theta}_2 + g\sin\theta_2 = 0 \tag{4.30}$$

respectively. Clearly, $\theta = \theta_1 + \theta_2$ is NOT a solution of Eq. (4.28), since

$$L(\ddot{\theta}_1 + \ddot{\theta}_2) + g\sin(\theta_1 + \theta_2) \neq L\ddot{\theta}_1 + g\sin\theta_1 + L\ddot{\theta}_2 + g\sin\theta_2 \tag{4.31}$$

However, if the displacement angle θ is small and the small vibration assumption can be used, Eq. (4.28) will be linear. For small angle, we have

$$\sin\theta \simeq \theta \tag{4.32}$$

and we obtain the linear system

$$L\ddot{\theta} + g\theta = 0 \tag{4.33}$$

Clearly, if we are interested in the analysis of the pendulum for large amplitude of vibrations, we must use Eq. (4.28), rather than Eq. (4.33). It requires a nonlinear treatment which is beyond the scope of this textbook.

In this textbook, only the linear vibration system will be discussed. If two impulses occur at two different times, their responses will superimpose.

Example 4.2

For a linear system shown in Fig. 4.4, assume that $m = 1$ kg; $c = 0.5$ kg/s; $k = 4$ N/m; $f_1 = f_2 = 1$ Ns, what is the response when two impulses are applied 5 seconds apart?

Solution:

First, we can obtain the undamped natural frequency and the damping ratio as follows:

$$\omega_n = \sqrt{\frac{k}{m}} = \sqrt{\frac{4}{1}} \text{ rad/s} = 2 \text{ rad/s}$$

$$\zeta = \frac{c}{2\sqrt{km}} = \frac{0.5}{2\sqrt{4}} = 0.125$$

By using Eq. (2.75), we can obtain the damped natural frequency

$$\omega_d = \omega_n \sqrt{1-\zeta^2} = 2\sqrt{1-0.125^2} \text{ rad/s} = 1.984 \text{ rad/s}$$

The solution can be written as (based on Eq. (4.20))

$$h_1(t) = \frac{\exp(-0.25t)}{1.984}\sin(1.984t), \quad t > 0 \tag{4.34}$$

$$h_2(t) = \frac{\exp[-0.25(t-5)]}{1.984}\sin[1.984(t-5)], \quad t > 5 \tag{4.35}$$

and the total solution is

$$h_T(t) = \begin{cases} \dfrac{\exp(-0.25t)}{1.984}\sin(1.984t) & 0 < t < 5 \\ \dfrac{\exp(-0.25t)}{1.984}\sin(1.984t) + \dfrac{\exp[-0.25(t-5)]}{1.984}\sin[1.984(t-5)] & t \geqslant 5 \end{cases} \tag{4.36}$$

Eqs. (4.34)—(4.36) can be plotted as in Fig. 4.6. Fig. 4.6(a) and (b) show the results for Eqs. (4.34) and (4.35), respectively. Fig. 4.6(c) shows the complete results by using the principle of superposition.

4.3 Response of SDOF Systems under a General Periodic Force

If the forcingfunction $F(t)$ is periodic, we can use the Fourier series and the principle of superposition to get the response. The Fourier series states that a periodic function can be represented as a series of sines and cosines:

$$\begin{aligned} F(t) &= \frac{a_0}{2} + a_1\cos(\omega t) + a_2\cos(2\omega t) + \cdots + b_1\sin(\omega t) + b_2\sin(2\omega t) + \cdots \\ &= \frac{a_0}{2} + \sum_{j=1}^{\infty}[a_j\cos(j\omega t) + b_j\sin(j\omega t)] \end{aligned} \tag{4.37}$$

where

$$a_j = \frac{2}{T}\int_0^T F(t)\cos(j\omega t)\mathrm{d}t, \quad j = 0, 1, 2, \cdots$$

$$b_j = \frac{2}{T}\int_0^T F(t)\sin(j\omega t)\mathrm{d}t, \quad j = 1, 2, \cdots$$

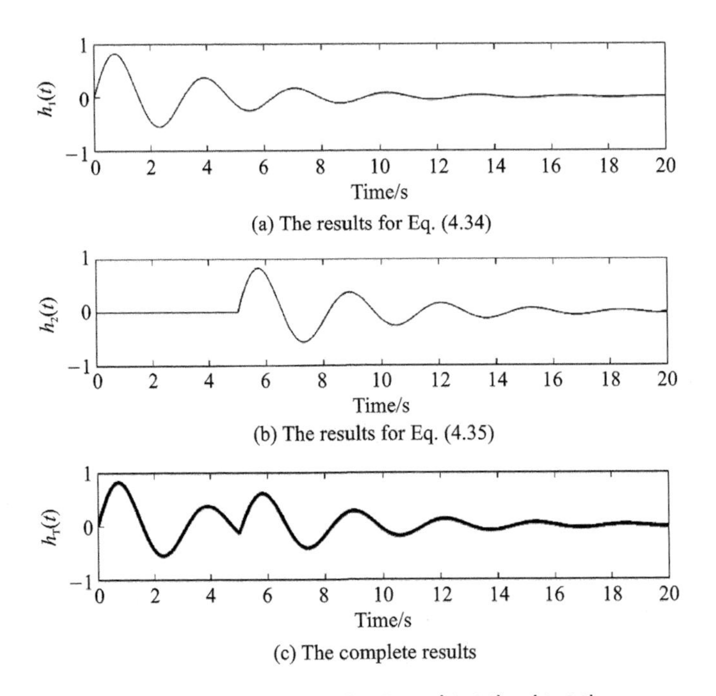

Fig. 4.6 The results for Eqs. (4.34)—(4.36)

and $T = 2\pi/\omega$ is the period. If a mass-spring-damper system is excited by a periodic force $F(t)$, the equation of motion can be written as

$$m\ddot{x}+c\dot{x}+kx=F(t)=\frac{a_0}{2}+\sum_{n=1}^{\infty}[a_n\cos(n\omega t)+b_n\sin(n\omega t)] \tag{4.38}$$

Using the principle of superposition, the steady-state solution of this equation is the sum of the steady-state solutions of each part, such as

$$m\ddot{x}+c\dot{x}+kx=\frac{a_0}{2} \tag{4.39a}$$

$$m\ddot{x}+c\dot{x}+kx=a_n\cos(n\omega t) \tag{4.39b}$$

$$m\ddot{x}+c\dot{x}+kx=b_n\sin(n\omega t) \tag{4.39c}$$

The particular solution for Eq. (4.39a) can be easy to obtained as follows:

$$x_p(t)=\frac{a_0}{2k} \tag{4.40}$$

According to Chapter 3, it is easy to find that the particular solutions for Eqs. (4.39b) and (4.39c) are

$$x_p(t)=\frac{a_n}{k}\frac{1}{\sqrt{(1-k^2\beta^2)^2+(2\zeta n\beta)^2}}\cos(n\omega t-\phi_n) \tag{4.41}$$

$$x_p(t)=\frac{b_n}{k}\frac{1}{\sqrt{(1-k^2\beta^2)^2+(2\zeta n\beta)^2}}\sin(n\omega t-\phi_n) \tag{4.42}$$

$$\phi_n = \arctan \frac{2\zeta n\beta}{1 - k^2\beta^2} \tag{4.43}$$

Then add up all the sums to get the complete steady-state solution, such as

$$x_p(t) = \frac{a_0}{2k} + \sum_{n=1}^{\infty} \frac{a_n}{k} \frac{1}{\sqrt{(1-k^2\beta^2)^2 + (2\zeta n\beta)^2}} \cos(n\omega t - \phi_n) + \sum_{n=1}^{\infty} \frac{b_n}{k} \frac{1}{\sqrt{(1-k^2\beta^2)^2 + (2\zeta n\beta)^2}} \sin(n\omega t - \phi_n) \tag{4.44}$$

From Eq. (4.44), it can be found that the response will be significantly large if $n\omega = \omega_n$, especially for small n and ζ. Furthermore, the term $(2\zeta n\beta)^2$ will be increased as n becomes large. It means that the response becomes smaller and the corresponding terms tend to zero. So the next question is how many terms do we need to include?

Example 4.3

Obtain the steady-state response of a mass-spring-damper system having $m = 1$, $c = 0.5$ and $k = 1$ when subjected to the force shown in Fig. 4.7.

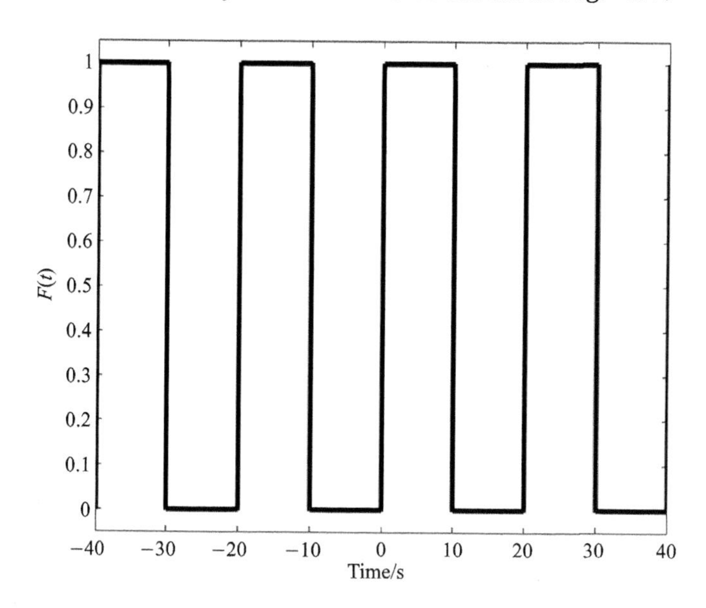

Fig. 4.7 A square input waveform

Solution:

From Fig. 4.7, it can be found that

$$T = 20, \quad \omega = \frac{2\pi}{T} = \frac{\pi}{10} \tag{4.45}$$

And the applied force function can be written as

$$F(t)=\begin{cases}1 & 0\leqslant t+nT\leqslant 10\\ -1 & 10\leqslant t+nT\leqslant 20\end{cases}\quad n=-\infty,\cdots,-1,0,1,\cdots,+\infty \tag{4.46}$$

The Fourier series of the force function in Eq. (4.46) can be written as

$$F(t)=\frac{a_0}{2}+\sum_{j=1}^{\infty}[a_j\cos(j\omega t)+b_j\sin(j\omega t)] \tag{4.47}$$

From Eq. (4.37), it can be found that

$$a_0=\frac{2}{T}\int_0^T F(t)\mathrm{d}t=\frac{2}{20}\left(\int_0^{10}\mathrm{d}t-\int_{10}^{20}\mathrm{d}t\right)=0.1[10-(20-10)]=0 \tag{4.48}$$

$$b_j=\frac{2}{T}\int_0^T F(t)\sin(j\omega t)\mathrm{d}t \tag{4.49}$$

$$\begin{aligned}a_n&=\frac{2}{T}\int_0^T F(t)\cos(n\omega t)\mathrm{d}t\\&=\frac{2}{20}\left[\int_0^{10}\cos\left(\frac{n\pi}{10}t\right)\mathrm{d}t-\int_{10}^{20}\cos\left(\frac{n\pi}{10}t\right)\mathrm{d}t\right]\\&=0.1\left[\frac{10}{n\pi}\cdot\sin\left(\frac{n\pi}{10}t\right)\Bigg|_{t=0}^{10}-\frac{10}{n\pi}\cdot\sin\left(\frac{n\pi}{10}t\right)\Bigg|_{t=10}^{20}\right]\\&=0\end{aligned} \tag{4.50}$$

From Eqs. (4.48) and (4.50), it can be found that all cosine terms vanish. From Eq. (4.49), it can be found that

$$\begin{aligned}b_n&=\frac{2}{T}\int_0^T F(t)\sin(n\omega t)\mathrm{d}t\\&=\frac{2}{20}\left[\int_0^{10}\sin\left(\frac{n\pi}{10}t\right)\mathrm{d}t-\int_{10}^{20}\sin\left(\frac{n\pi}{10}t\right)\mathrm{d}t\right]\\&=0.1\left[\frac{-10}{n\pi}\cdot\cos\left(\frac{n\pi}{10}t\right)\Bigg|_{t=0}^{10}-\frac{-10}{n\pi}\cdot\cos\left(\frac{n\pi}{10}t\right)\Bigg|_{t=10}^{20}\right]\\&=0.1\left\{\frac{-10}{n\pi}\cdot[\cos(n\pi)-1]+\frac{-10}{n\pi}\cdot[\cos(2n\pi)-\cos(n\pi)]\right\}\\&=\begin{cases}\frac{4}{n\pi} & n=\text{odd}\\ 0 & n=\text{even}\end{cases}\end{aligned} \tag{4.51}$$

In this way, the forcein Fig. 4.7 can be represented by a Fourier series as

$$\begin{aligned}F(t)&=b_1\sin(\omega t)+b_3\sin(3\omega t)+b_5\sin(5\omega t)+\cdots\\&=\sum_{n=1}^{\infty}b_n\sin(n\omega t)=\sum_{n=1}^{\infty}\frac{4}{n\pi}\sin\left(n\frac{\pi}{10}t\right),\quad n=1,3,5,\cdots\end{aligned} \tag{4.52}$$

Eq. (4.52) can be plotted graphically as in Fig. 4.8.

And the following MATLAB program can be used to calculate Eq. (4.52), and the results are shown in Fig. 4.8.

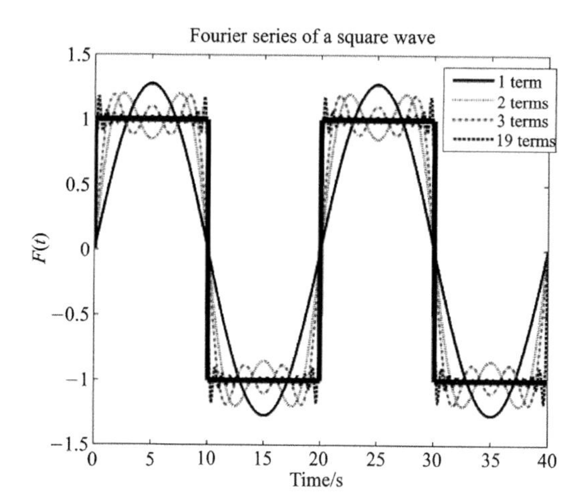

Fig. 4.8 Fourier series of a square wave

MATLAB Codes	Comments
clear; close all;	Remove all variables in workspace and close all figures
t=linspace(0,40,1e3);	Generates a row vectort of 1 000 points linearly spaced from 0 to 40
as=0;	The terms
for n=1:2:30	n= 1,3,5,7, …
as=as+1;	Each loop to add one term
s(as,:)=4/n/pi * sin(n * pi * t/10);	Calculate Eq. (4.52)
if as==1 S1=s; end	Using one term
if as==2 S2=sum(s); end	Using two terms
if as==3 S3=sum(s); end	Using three terms
if as==15 S4=sum(s); end	Using fifteen terms
end	End the loop
figure(1),	Create a new figure
plot(t,S1,'k','linewidth',2), hold on	Plot the S1 by solid line (black color and linewidth=2 points)
plot(t,S2,'b:','linewidth',2)	Plot the S2
plot(t,S3,'g-.','linewidth',2)	Plot the S3
plot(t,S4,'r--','linewidth',2)	Plot the S4
w=pi/10;plot(t,sign(sin(w * t)),'b', 'linewidth',4)	Plot the square wave
legend('1 term','2 terms','3 terms','19 terms')	Display a legend on the figure
xlabel('Time (s)','fontsize',12);ylabel ('F(t)','fontsize',12)	Label the x-axis and y-axis
title('Fourier series of a square wave')	Add a title for the figure

By using Eqs. (4.44) and (4.52), the response of the dynamic system with a square input can be obtained. Notice that it is tedious to solve Eq. (4.52) by hand. So the following MATLAB program is used to calculate the response, and Fig. 4.9 shows the results under different terms of Fourier series. Fig. 4.10 shows the contribution of each term to total response.

```
% ***********************************************************
% Periodic response of a dynamic system to a square input waveform
% ***********************************************************
clear; closeall;
m=1; c=0.5; k=1;% Parameters
wn=sqrt(k/m);
zi=c/wn/2;
w=pi/10;
t=linspace(0,40,1e3);
x=zeros(size(t));
r=w/wn;
for jj=1:2:19
a(jj)=4/pi/jj;
X(jj) = a (jj)/k/sqrt ((1 - jj^2 * r^2)^2 + (2 * zi * jj * r)^2);% term
in summation
phi(jj)=atan2(2 * zi * jj * r,1-jj * jj * r * r);
x(jj+2,:)=x(jj,:)+X(jj) * sin(w * jj * t-phi(jj));
end
u=sign(sin(w * t));
figure(2)
%plot(t,x([3 5 7 21],:),t,sign(sin(w * t)),'k','linewidth',2);
plot(t,x(3,:),'k','linewidth',2);
holdon
plot(t,x(5,:),'k:','linewidth',2);
plot(t,x(7,:),'k-.','linewidth',2);
plot(t,x(19,:),'b','linewidth',4);
grid
legend('1 term','2 terms','3 terms','19 terms')
xlabel('Time (s)','fontsize',12);ylabel('x(t)','fontsize',12)
title('Response to periodic input','fontsize',12);
figure(3)
```

```
plot(1:2:19,X(1:2:19),'o','markersize',10,'linewidth',4);
grid
ylabel('Magnitude of term in summation','fontsize',12)
xlabel('Terms','fontsize',12)
```

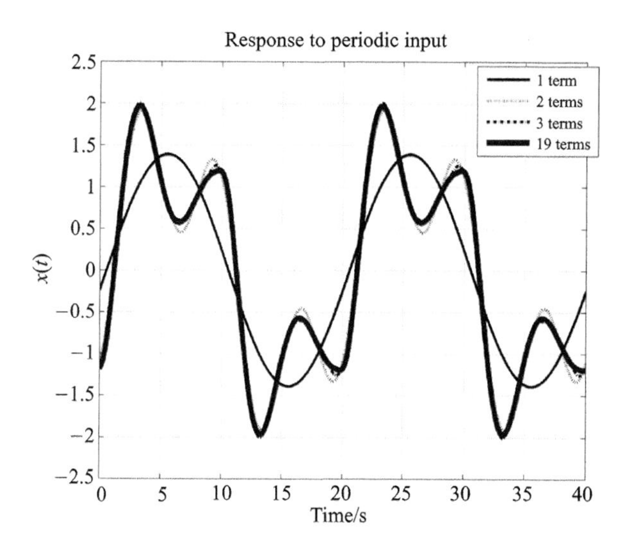

Fig. 4.9 The response under different terms of Fourier series

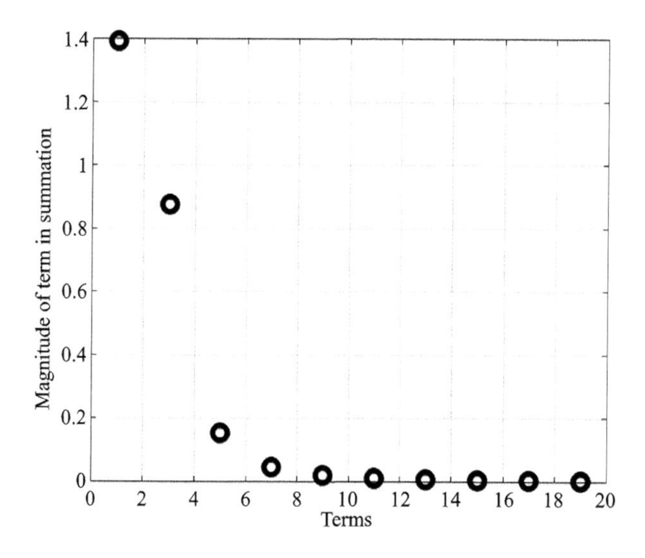

Fig. 4.10 The contribution of each term to total response

4.4 System's Response to General Excitation by Convolution

We now focus our attention on the general excitation by the force $f(t)$ of the mass-spring-damper shown in Fig. 4.11. In this case, the SDOF system is excited by a general excitation force. For this the convolution integral is imposed.

An arbitrary function $f(t)$ can be regarded as an impulse of magnitude $f(h)\Delta\tau$ if the time interval $d\tau$ is small enough, as shown in Fig. 4.12. Thus we get

$$df(t,\tau)=f(\tau)d\tau\delta(t-\tau)=f(\tau)\delta(t-\tau)d\tau \tag{4.53}$$

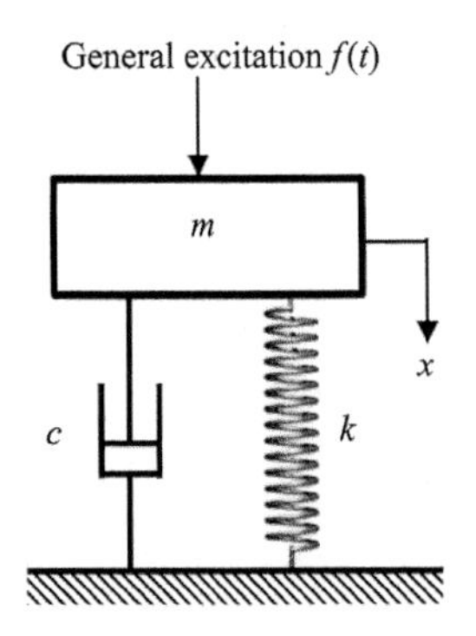

Fig. 4.11 A SDOF system under general excitation

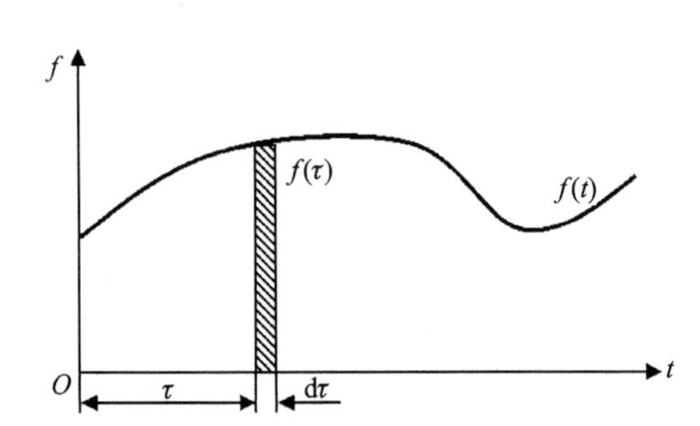

Fig. 4.12 An arbitrary function $f(t)$

By using Eq. (4.53), the response of the system to the excitation at $t=\tau$ is thus

$$dx(t,\tau)=f(\tau)h(t-\tau)d\tau \tag{4.54}$$

Clearly, when $d\tau\to 0$, the response to the total function $x(t)$ can be obtained by using the convolution integral

$$x(t)=\int_0^t f(\tau)h(t-\tau)d\tau \tag{4.55}$$

Since τ is a dummy variable of integration, we may write Eq. (4.55) in the following equivalent form

$$x(t)=\int_0^t f(t-\tau)h(\tau)d\tau \tag{4.56}$$

Note: This equation does not consider initial conditions. This type of formula is called the convolution integral or the Duhamel's integral.

For the complete response of the mass-spring-damper system, the response due to the initial conditions should be included, such as

$$x(t)=\exp(-\zeta\omega_n t)[A_1\cos(\omega_d t)+A_2\sin(\omega_d t)]+\int_0^t f(t-\tau)h(\tau)d\tau \tag{4.57}$$

where $A_1 = x_0$, $A_2 = \dfrac{v_0 + \zeta\omega_n x_0}{\omega_d}$, ζ is the damping ratio, and ω_d is the damped natural frequency.

Example 4.4

The system shown in Fig. 4.11 is initially at rest. At the time $t = 0$, a constant force F_0 is applied on the system. Assume that damper $c = 0$, determine the response of the system.

Solution:

The differential equation of motioncan be written as

$$m\ddot{x} + kx = F_0 \tag{4.58}$$

and the initial conditions are

$$\dot{x}(0) = x(0) = 0 \tag{4.59}$$

Obviously, for this undamped system, $\zeta = 0$ and hence $\omega_d = \omega_n$. From Eq. (4.20), we have

$$h(t-\tau) = \frac{1}{m\omega_n}\sin[\omega_n(t-\tau)] \tag{4.60}$$

By using the convolution integral in Eq. (4.56), it gives

$$\begin{aligned} x(t) &= \int_0^t \frac{F_0}{m\omega_n}\sin[\omega_n(t-\tau)]\,d\tau = \frac{F_0}{m\omega_n}\frac{1}{\omega_n}\cos[\omega_n(t-\tau)]\Big|_{\tau=0}^{t} \\ &= \frac{F_0}{m\omega_n^2}\{\cos[\omega_n(t-t)] - \cos(\omega_n t)\} = \frac{F_0}{m\omega_n^2}[1-\cos(\omega_n t)] \end{aligned} \tag{4.61}$$

Notice that natural frequency $\omega_n^2 = \dfrac{k}{m}$, the solution can be rewritten as

$$x(t) = \frac{F_0}{k}[1-\cos(\omega_n t)] \tag{4.62}$$

Example 4.5

In Example 4.4, if damper $c > 0$, determine the response of the system.

Solution:

In this case, the differential equation of motion can be expressed as

$$m\ddot{x} + c\dot{x} + kx = F_0 \tag{4.63}$$

and the initial conditions are

$$\dot{x}(0) = x(0) = 0 \tag{4.64}$$

By using Eqs. (4.20) and (4.56), we have

$$\begin{aligned} x(t) &= \int_0^t F_0 h(t-\tau)\,d\tau \\ &= \int_0^t \frac{F_0}{m\omega_d}\exp[-\zeta\omega_n(t-\tau)]\sin[\omega_d(t-\tau)]\,d\tau \end{aligned}$$

$$=\frac{F_0}{k}\left\{1-\frac{\exp(-\zeta\omega_n t)}{\sqrt{1-\zeta^2}}\cos\left[\omega_d(t-\phi)\right]\right\} \tag{4.65}$$

where $\phi=\arctan\frac{\zeta}{\sqrt{1-\zeta^2}}$.

Clearly, if $\zeta=0$, $x(t)=\frac{F_0}{k}[1-\cos(\omega_n t)]=\frac{F_0}{m\omega_n^2}[1-\cos(\omega_n t)]$, it is the same to solution in Example 4.4. Fig. 4.13 shows the responses under different damping ratios by using Eq. (4.65). It can be found that the maximum dynamic displacement is twice the static displacement occurring under the same load. This is an important consideration in structures subjected to suddenly applied loads.

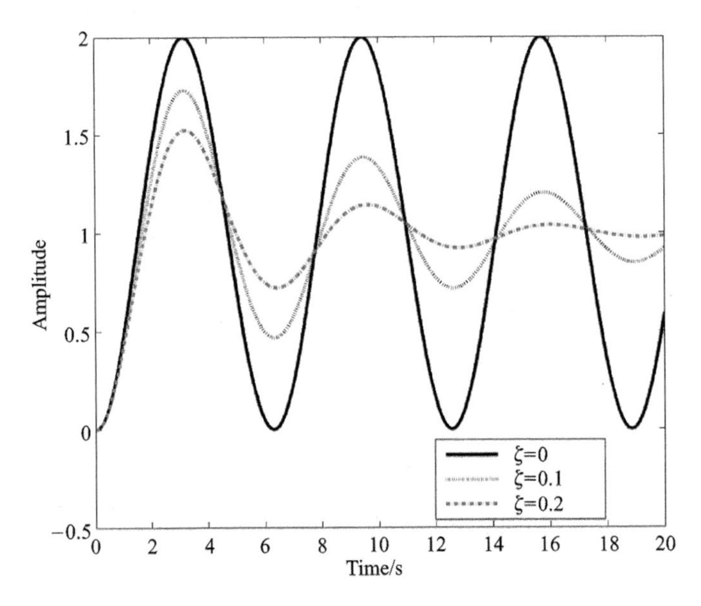

Fig. 4.13 The responses of the mass-spring-damper system under different damping ratios

Example 4.6

The undamped SDOF system (damping $c=0$) shown in Fig. 4.14 is initially at rest. At the time $t=0$ a harmonic force $f(t)=F_0\sin(\omega t)$ is applied on the system. Determine the response of the system.

Solution:

Notice that the damping c is zero, so the differential equation of motion can be expressed as

$$m\ddot{x}+kx=F_0\sin(\omega t),\quad t>0 \tag{4.66}$$

and the initial conditions are

$$\dot{x}(0)=x(0)=0 \tag{4.67}$$

By using Eq. (4.20), the impulse response can be written as

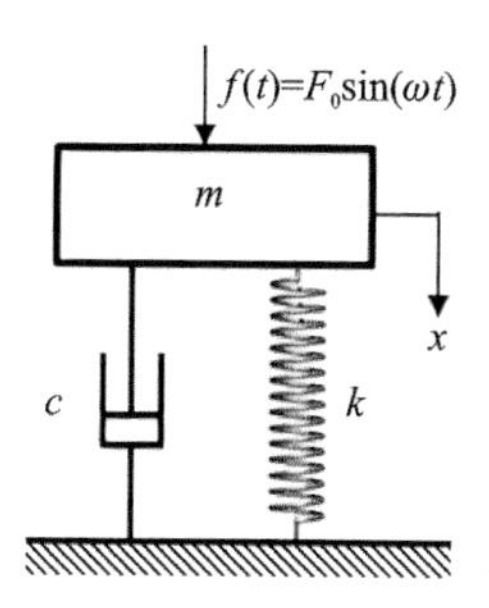

Fig. 4.14 A harmonic force $f(t)=F_0\sin(\omega t)$ is applied on the system at the time $t=0$

$$h(t-\tau)=\frac{1}{m\omega_n}\sin[\omega_n(t-\tau)] \tag{4.68}$$

and the convolution integral in Eq. (4.56) is given by

$$x(t)=\int_0^t \frac{F_0\sin(\omega\tau)}{m\omega_n}\sin[\omega_n(t-\tau)]\,d\tau \tag{4.69}$$

Using the trigonometric relation

$$\sin\alpha\sin\beta=\frac{1}{2}[\cos(\alpha-\beta)-\cos(\alpha+\beta)] \tag{4.70}$$

We have

$$x(t)=\frac{F_0}{2m\omega_n}\int_0^t\{\cos[(\omega+\omega_n)\tau-\omega_n t]-\cos[(\omega-\omega_n)\tau+\omega_n t]\}\,d\tau \tag{4.71}$$

From Eqs. (4.70) and (4.71), we get

$$\begin{aligned} x(t)&=\frac{F_0}{2m\omega_n}\left\{\frac{\sin[(\omega+\omega_n)\tau-\omega_n t]}{\omega+\omega_n}-\frac{\sin[(\omega-\omega_n)\tau+\omega_n t]}{\omega-\omega_n}\right\}\Bigg|_{\tau=0}^{t} \\ &=\frac{F_0}{2m\omega_n}\left[\frac{\sin(\omega t)}{\omega+\omega_n}-\frac{\sin(\omega t)}{\omega-\omega_n}+\frac{\sin(\omega_n t)}{\omega+\omega_n}+\frac{\sin(\omega_n t)}{\omega-\omega_n}\right] \end{aligned} \tag{4.72}$$

Notice that $\omega_n^2=\frac{k}{m}$, Eq. (4.72) can be rewritten as

$$x(t)=\frac{F_0}{k\left[1-\left(\frac{\omega}{\omega_n}\right)^2\right]}\left[\sin(\omega t)-\frac{\omega}{\omega_n}\sin(\omega_n t)\right] \tag{4.73}$$

An alternative approach: This problem may also be solved alternatively as follows. Substituting $\beta=\frac{\omega}{\omega_n}$, $A=x_0$ and $B=\frac{v_0}{\omega_n}$ in Eq. (3.39) in Chapter 3, it can be found that the general solution of Eq. (4.66) is given by

$$x(t)=\frac{F_0\sin(\omega t)}{k\left[1-\left(\frac{\omega}{\omega_n}\right)^2\right]}+x_0\cos(\omega_n t)+\frac{v_0}{\omega_n}\sin(\omega_n t) \tag{4.74}$$

where x_0 and v_0 may be determined uniquely by the initial conditions. The initial condition $x(0)=0$ gives $x_0=0$. Hence

$$x(t)=\frac{F_0\sin(\omega t)}{k\left[1-\left(\frac{\omega}{\omega_n}\right)^2\right]}+\frac{v_0}{\omega_n}\sin(\omega_n t) \tag{4.75}$$

Differentiating Eq. (4.75) with the time t, we have

$$\dot{x}(t)=\frac{F_0\omega\cos(\omega t)}{k\left[1-\left(\frac{\omega}{\omega_n}\right)^2\right]}+v_0\cos(\omega_n t) \tag{4.76}$$

Therefore, the initial velocity can be written as

$$\dot{x}(0)=\frac{F_0\omega}{k\left[1-\left(\frac{\omega}{\omega_n}\right)^2\right]}+v_0=0 \tag{4.77}$$

From Eq. (4.77), it is easy to find that

$$v_0=\frac{-F_0\omega}{k\left[1-\left(\frac{\omega}{\omega_n}\right)^2\right]} \tag{4.78}$$

Substituting Eq. (4.78) and $x_0=0$ into Eq. (4.74), the response shown in Eq. (4.73) is obtained.

4.5 System's Response to General Excitation by the Laplace Transform

The Laplace transform $X(s)$ of a function $x(t)$ is defined by

$$X(s)=\mathscr{L}[x(t)]=\int_0^{\infty}x(t)\exp(-st)\mathrm{d}t \tag{4.79}$$

where $s=\sigma+\mathrm{j}\omega$, and the real numbers σ and ω denote the real and imaginary parts of s, respectively. $\mathscr{L}[\,]$ is Laplace operator.

Recalling the general differential relation (from the theory ofordinary differential equations)

$$\mathscr{L}\left[\frac{\mathrm{d}^n x(t)}{\mathrm{d}t^n}\right]=s^n X(s)-s^{n-1}x(0)-s^{n-2}\frac{\mathrm{d}x(0)}{\mathrm{d}t}-\cdots-\frac{\mathrm{d}^{n-1}x(0)}{\mathrm{d}t^{n-1}} \tag{4.80}$$

If we set $n=1$, it can be found that

$$\mathscr{L}[\dot{x}(t)]=sX(s)-x(0) \tag{4.81}$$

and if setting $n=2$,

$$\mathscr{L}[\ddot{x}(t)]=s^2X(s)-sx(0)-\dot{x}(0) \tag{4.82}$$

So, using Eqs. (4.81) and (4.82), the Laplace transform of our mass-spring-damper system

$$\left.\begin{aligned}&m\ddot{x}+c\dot{x}+kx=f(t)\\&x(0)=x_0,\dot{x}(0)=v_0\end{aligned}\right\} \tag{4.83}$$

can be expressed as

$$m\left[s^2X(s)-sx_0-v_0\right]+c\left[sX(s)-x_0\right]+kX(s)=F(s) \tag{4.84}$$

where $F(s)$ is the Laplace transform of the excitation $f(t)$. We may rewrite Eq. (4.84) in the form

$$X(s)=\frac{F(s)}{ms^2+cs+k}+\frac{(ms+c)x_0+mv_0}{ms^2+cs+k} \tag{4.85}$$

In practice $F(s)$ is usually determined from a table of Laplace transform (in Appendix C). Then $X(s)$ can be obtained by Eq. (4.85). Finally the response $x(t)$, called the inverse Laplace transform of $X(s)$, is determined from the table of the transformation.

The inverse Laplace transform of $X(s)$ in Eq. (4.85) can be expressed as

$$x(t)=x_0\exp(-\zeta\omega_n t)\cos(\omega_d t)+\frac{v_0+\zeta\omega_n x_0}{\omega_d}\exp(-\zeta\omega_n t)\sin(\omega_d t)+$$
$$\frac{1}{m\omega_d}\int_0^t \exp\left[-\zeta\omega_n(t-\tau)\right]\sin\left[\omega_d(t-\tau)\right]f(\tau)\mathrm{d}\tau \tag{4.86}$$

The following example demonstrates the use of this method.

Example 4.7

An undamped SDOF system is shown in Fig. 4.15, assume that a mass is dropped from height H. At $t=0$ it touches the spring with stiffness k. If $x(t)$ is the displacement of the mass, measured from undeformed position of the spring at $t=0$ (ie. $x(0)=0$), determine the motion of the mass.

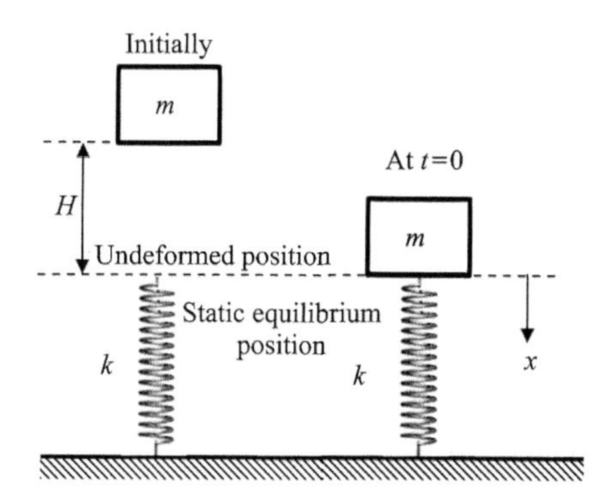

Fig. 4.15 A mass dropped from height H

Solution:

Clearly, there is no damping in this example, such as $c=0$. The motion equation can be expressed as

$$m\ddot{x}+kx=mg,\quad t>0 \tag{4.87}$$

We can set the initial position as just before the spring touches the spring, so the initial conditions can be expressed as

$$x(0)=0,\quad \dot{x}(0)=\sqrt{2gH} \tag{4.88}$$

By using Eq. (4.85), we get

$$X(s)=\frac{F(s)}{ms^2+k}+\frac{m\sqrt{2gH}}{ms^2+k} \tag{4.89}$$

The Laplace transform of 1 is $1/s$ (see Laplace transform pairs listed in Appendix C). Therefore

$$F(s)=\mathscr{L}(mg)=\frac{mg}{s} \tag{4.90}$$

Substituting Eq. (4.90) into Eq. (4.89), we get

$$X(s)=\frac{mg}{(ms^2+k)s}+\frac{m\sqrt{2gH}}{ms^2+k} \tag{4.91}$$

By using the table of Laplace transform pairs listed in Appendix C

$$f(t)=\frac{1-\cos(\omega t)}{\omega^2}\Leftrightarrow\mathscr{L}[f(t)]=\frac{1}{(s^2+\omega^2)s} \tag{4.92}$$

We therefore have

$$\mathscr{L}^{-1}\left[\frac{mg}{s(ms^2+k)}\right]=\mathscr{L}^{-1}\left[\frac{g}{s(s^2+\omega_n^2)}\right]=\frac{g}{\omega_n^2}[1-\cos(\omega_n t)] \tag{4.93}$$

By using the table of Laplace transform pairs listed in Appendix C

$$f(t)=\sin(\omega t)\Leftrightarrow\mathscr{L}[f(t)]=\frac{\omega}{s^2+\omega^2} \tag{4.94}$$

hence

$$\mathscr{L}^{-1}\left[\frac{m\sqrt{2gH}}{ms^2+k}\right]=\mathscr{L}^{-1}\left[\frac{\sqrt{2gH}}{\omega_n}\frac{\omega_n}{s^2+\omega_n^2}\right]=\frac{\sqrt{2gH}}{\omega_n}\sin(\omega_n t) \tag{4.95}$$

Using Eqs. (4.93) and (4.95), we find that the inverse Laplace transform of $X(s)$, given by Eq. (4.93), is

$$x(t)=\frac{g}{\omega_n^2}[1-\cos(\omega_n t)]+\frac{\sqrt{2gH}}{\omega_n}\sin(\omega_n t) \tag{4.96}$$

It should be noticed that Example 4.7 can be solved by more simple method by setting the coordinate origin at static equilibrium position, as shown in the following example.

Example 4.8 (Example 4.7 revisited)

In Example 4.7, notice that there is no force applied after the mass touches the spring. So it can be seen as a free vibration problem, and we set the coordinate origin at static equilibrium position, $x'=x+\Delta$, where Δ is the static deformation of the spring, and $\Delta=mg/k$, as shown in Fig. 4.16. So the solution $x'(t)$ can be expressed as

$$m\ddot{x}'+kx'=0 \tag{4.97}$$

with initial conditions

$$x(0)=-\frac{mg}{k},\quad \dot{x}(0)=\sqrt{2gH} \tag{4.98}$$

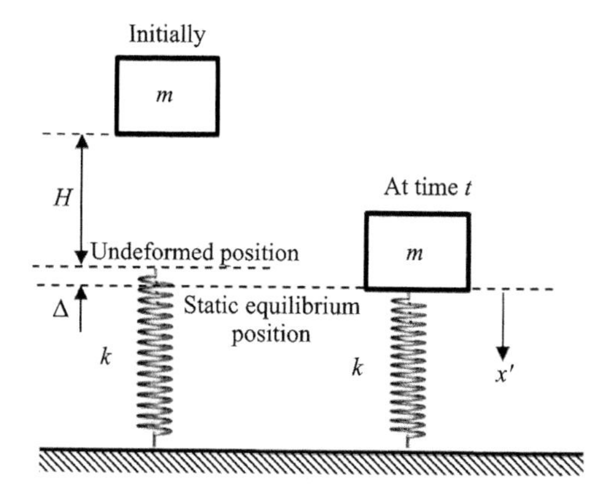

Fig. 4.16 A mass dropped from height *H* (free vibration problem)

Clearly, the solution for the free vibration of the SDOF system can be expressed as

$$x'(t)=x_0\cos(\omega_n t)+\frac{v_0}{\omega_n}\sin(\omega_n t) \tag{4.99}$$

Substituting Eq. (4.98) into Eq. (4.99), we get

$$x'(t)=-\frac{mg}{k}\cos(\omega_n t)+\frac{\sqrt{2gH}}{\omega_n}\sin(\omega_n t) \tag{4.100}$$

If we want to measure from undeformed position of the spring,

$$x(t)=x'(t)+\Delta=\frac{mg}{k}\left[1-\cos(\omega_n t)\right]+\frac{\sqrt{2gH}}{\omega_n}\sin(\omega_n t) \tag{4.101}$$

Clearly, Eq. (4.101) is the same as obtained in Eq. (4.96) by the method of Laplace transform.

Another alternative solution may be obtained by breaking Eq. (4.87) down into the following two simpler problems, such as

$$\left.\begin{aligned} m\ddot{x}_1+kx_1&=mg \\ x_1(0)=0,\ \dot{x}_1(0)&=0 \end{aligned}\right\} \tag{4.102a}$$

$$\left.\begin{aligned} m\ddot{x}_2+kx_2&=0 \\ x_2(0)=0,\ \dot{x}_2(0)&=\sqrt{2gH} \end{aligned}\right\} \tag{4.102b}$$

and then superimposing the solutions

$$x(t)=x_1(t)+x_2(t) \tag{4.103}$$

It is easily shown that Eq. (4.103) is indeed the solution of Eq. (4.87). Adding Eqs. (4.102a) and (4.102b) gives

$$\left.\begin{array}{l}(m\ddot{x}_1+kx_1)+(m\ddot{x}_2+kx_2)=mg+0=mg\\ x_1(0)+x_2(0)=0,\ \dot{x}_1(0)+\dot{x}_2(0)=\sqrt{2gH}\end{array}\right\} \tag{4.104}$$

Rearranging Eq. (4.104) and using Eq. (4.103), we find that $x(t)$ given by Eq. (4.103) satisfies Eq. (4.87). The solution for Eq. (4.102a) is given by Eq. (4.65) (with $F_0=mg$) :

$$x_1(t)=\frac{mg}{k}[1-\cos(\omega_n t)] \tag{4.105}$$

and the solution for Eq. (4.102b) is given by Eq. (4.19) (with $x_0=0$, $v_0=2gH$):

$$x_2(t)=\frac{\sqrt{2gH}}{\omega_n}\sin(\omega_n t) \tag{4.106}$$

Hence, substituting Eqs. (4.105) and (4.106) into Eq. (4.103), the solution $x(t)$ for Eq. (4.99) can be obtained:

$$x(t)=\frac{mg}{k}[1-\cos(\omega_n t)]+\frac{\sqrt{2gH}}{\omega_n}\sin(\omega_n t) \tag{4.107}$$

Clearly, Eq. (4.107) is also the same as obtained in Eq. (4.96) by the method of Laplace transform.

4.6 The Transfer Function

A system may be characterized by its differential equation of motion. In our case

$$m\ddot{x}+c\dot{x}+kx=f(t) \tag{4.108}$$

If we know m, c and k, then the system response $x(t)$ to the forcing function $f(t)$ can be determined be solving Eq. (4.108).

The system can alternatively be characterized by its impulse response. If we know the impulse response $h(t-\eta)$, then the response to the excitation $f(t)$ may be obtained by the convolution integral

$$x(t)=\int_0^t f(t)h(t-\eta)\mathrm{d}\eta \tag{4.109}$$

As noted above the input-output relation is determined in this case by the convolution in Eq. (4.109), or symbolically

$$x=f\otimes h \tag{4.110}$$

A third way to characterize the system is by its transfer function $H(s)$, formally defined as

$$H(s)=\frac{X(s)}{F(s)} \tag{4.111}$$

where $X(s)$ is the Laplace transform of the system response with zero initial conditions, and $F(s)$ is the Laplace transform of the exciting force. The input output relation in the Laplace domain is a simple multiplication

$$X(s)=H(s)F(s) \tag{4.112}$$

Fig. 4.17 shows the block diagram for a pure time domain analysis and Laplace domain analysis, respectively.

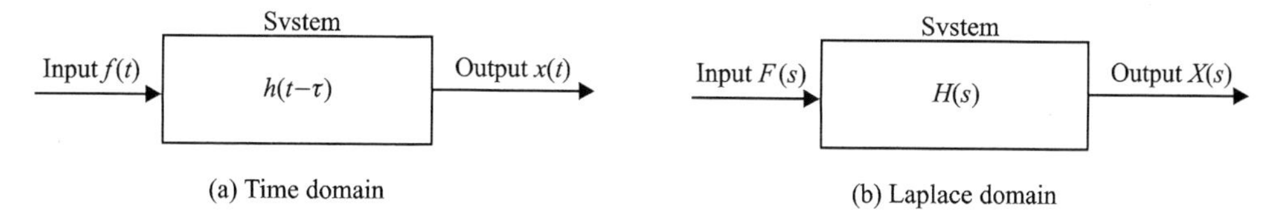

Fig. 4.17 A block diagram form for Time domain and Laplace transform

The use of Laplace transform is so popular because the multiplication in Eq. (4.112) is much easier to perform than the convolution of Eq. (4.110) in the time domain. This is based on the well-known convolution theorem that states: **The convolution of two time domain functions is equal to the inverse Laplace transformation of the product of their two transforms.**

4.7 Composite Function Excitation

A function that can be expressed by several, different, analytical expressions in some of its intervals of dependence is called a composite function. For example, the function shown in Fig. 4.18 is a composite function defined as

$$f(t)=\begin{cases} f_0\dfrac{t}{t_0} & 0<t\leqslant t_0 \\ f_0 & t_0<t\leqslant t_1 \\ f_0\dfrac{t_2-t}{t_2-t_1} & t_1<t\leqslant t_2 \end{cases} \tag{4.113}$$

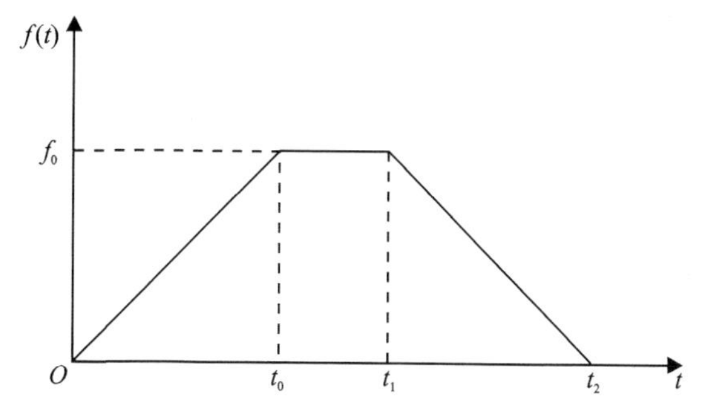

Fig. 4.18 A composite function

Consider the mass-spring-damper system

$$\left.\begin{array}{l} m\ddot{x}+c\dot{x}+kx=f(t) \\ x(0)=x_0,\dot{x}(0)=v_0 \end{array}\right\} \tag{4.114}$$

excited by the composite function

$$f(t)=\begin{cases} f_1(t) & 0<t\leqslant t_1 \\ f_1(t) & t_1<t\leqslant t_2 \\ \vdots & \vdots \\ f_n(t) & t_{n-1}<t\leqslant\infty \end{cases} \tag{4.115}$$

We may first find the response $x(t)$ for $0<t<t_1$. It is simply the solution of

$$\left.\begin{array}{l} m\ddot{x}+c\dot{x}+kx=f_1(t),0<t\leqslant t_1 \\ x(0)=x_0,\dot{x}(0)=v_0 \end{array}\right\} \tag{4.116}$$

If we know the initial conditions $x(0)$ and $\dot{x}(0)$, Eq. (4.116) can be solved. The solved $x(t)$ in Eq. (4.116) can be used as the initial conditions to the dynamic of the system in the second interval $t_1<t\leqslant t_2$, such as,

$$\left.\begin{array}{l} m\ddot{x}+c\dot{x}+kx=f_2(t),\ t_1<t\leqslant t_2 \\ x(t_1)\text{ and }\dot{x}(t_1)\text{ as obtained from Eq. (4.116)} \end{array}\right\} \tag{4.117}$$

Then the solution of Eq. (4.117) determines the initial conditions $x(t_2)$ and $\dot{x}(t_2)$ required for the formulation of the problem in the third time interval $t_2<t\leqslant t_3$. Continuing in this manner determines the response of Eq. (4.114) for all time t. The following example demonstrates the application of the method.

Example 4.9

Determine the response $x(t)$ of

$$\left.\begin{array}{l} m\ddot{x}+kx=f(t) \\ x(0)=0,\ \dot{x}(0)=0 \end{array}\right\} \tag{4.118}$$

where $f(t)$ is the step function as shown in Fig. 4.19:

$$f(t)=\begin{cases} F_0 & 0<t\leqslant t_1 \\ 0 & t>t_1 \end{cases} \tag{4.119}$$

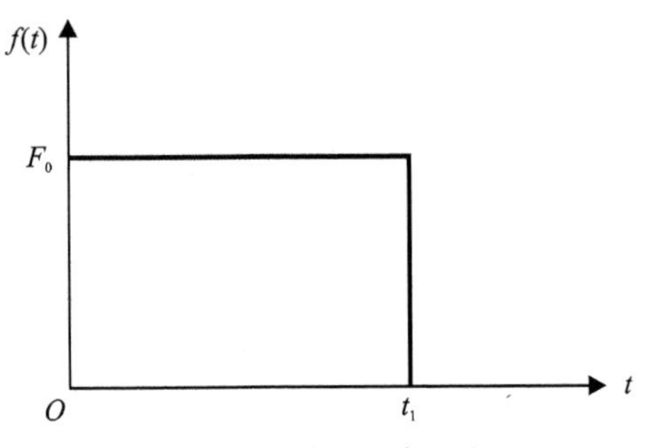

Fig. 4.19 A step function

Solution:

The response of the system during the time interval $0<t\leqslant t_1$ is determined by

$$\left.\begin{array}{l} m\ddot{x}+kx=F_0,0<t\leqslant t_1 \\ x(0)=0,\dot{x}(0)=0 \end{array}\right\} \tag{4.120}$$

It was found in Example 4.4 (see Eq. (4.62)) that the solution for Eq. (4.120) is

$$x(t)=\frac{F_0}{k}[1-\cos(\omega_n t)], \quad 0<t\leqslant t_1 \tag{4.121}$$

In particular we have

$$x(t_1)=\frac{F_0}{k}[1-\cos(\omega_n t_1)] \tag{4.122}$$

and

$$\dot{x}(t_1)=\frac{F_0\omega_n}{k}\sin(\omega_n t_1) \tag{4.123}$$

Hence, the response of the system for $t>t_1$ is governed by

$$\left.\begin{aligned} &m\ddot{x}+kx=0, t\geqslant t_1 \\ &x(t_1)=\frac{F_0}{k}[1-\cos(\omega_n t_1)] \\ &\dot{x}(t_1)=\frac{F_0\omega_n}{k}\sin(\omega_n t_1) \end{aligned}\right\} \tag{4.124}$$

It is always recommended to have the initial conditions expressed in terms of the time origin. We therefore introduce the time shift

$$\tau=t-t_1 \tag{4.125}$$

The system in Eq. (4.124) is thus

$$\left.\begin{aligned} &m\ddot{x}+kx=0, \tau\geqslant 0 \\ &x(0)=\frac{F_0}{k}[1-\cos(\omega_n t_1)] \\ &\dot{x}(0)=\frac{F_0\omega_n}{k}\sin(\omega_n t_1) \end{aligned}\right\} \tag{4.126}$$

According to Chapter 2, the solution for Eq. (4.126) can be expressed as

$$x=\frac{F_0}{k}[1-\cos(\omega_n t_1)]\cos(\omega_n\tau)+\frac{F_0}{k}\sin(\omega_n t_1)\sin(\omega_n\tau) \tag{4.127}$$

We may rewrite Eq. (4.127) in the form

$$x=\frac{F_0}{k}[\cos(\omega_n\tau)-\cos(\omega_n t_1)\cos(\omega_n\tau)+\sin(\omega_n t_1)\sin(\omega_n\tau)] \tag{4.128}$$

Hence by the trigonometric relation

$$\cos(\alpha+\beta)=\cos\alpha\cos\beta-\sin\alpha\sin\beta \tag{4.129}$$

Eq. (4.128) can be simplified as

$$x=\frac{F_0}{k}\{\cos(\omega_n\tau)-\cos[\omega_n(t_1+\tau)]\} \tag{4.130}$$

We now shift the time back, from τ to t. Using Eq. (4.125), we have

$$x=\frac{F_0}{k}\{\cos[\omega_n(t-t_1)]-\cos(\omega_n t)\}, \quad t\geqslant t_1 \tag{4.131}$$

In summary, by using Eqs. (4.122) and (4.131), the response of Eq.

(4.118) to the excitation in Eq. (4.119) is given by

$$x=\begin{cases}\dfrac{F_0}{k}[1-\cos(\omega_n t)] & 0<t\leqslant t_1\\ \dfrac{F_0}{k}\{\cos[\omega_n(t-t_1)]-\cos(\omega_n t)\} & t\geqslant t_1\end{cases} \tag{4.132}$$

It is worth while to understand the physical meaning of the solutionin Eq. (4.132). The dynamic of the system for this case can be regarded as the sum of two uniform excitations: one with constant amplitude F_0 starting at $t=0$, the other of amplitude $-F_0$ starting at $t=t_1$. This is demonstrated graphically in Fig. 4.20.

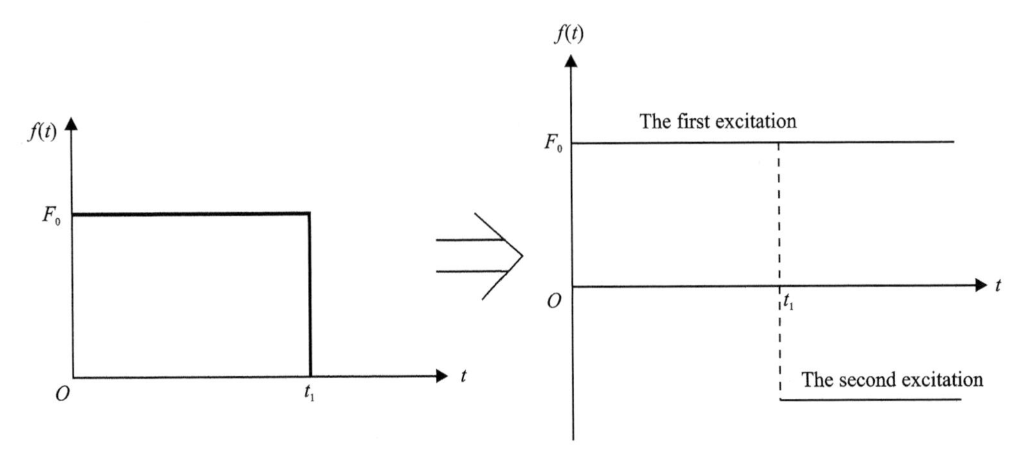

Fig. 4.20 The step force regarded as the sum of two uniform excitations

From Fig. 4.20, it can be found that the response $x_1(t)$ to the first excitation is (see Eq. (4.62) again)

$$x_1(t)=\frac{F_0}{k}[1-\cos(\omega_n t)],\quad 0<t<\infty \tag{4.133}$$

And the response $x_2(t)$ to the second excitation (by the same argument) is

$$x_2(t)=\begin{cases}0 & 0<t\leqslant t_1\\ -\dfrac{F_0}{k}\{1-\cos[\omega_n(t-t_1)]\} & t\geqslant t_1\end{cases} \tag{4.134}$$

The total response $x(t)=x_1(t)+x_2(t)$ is precisely the response given in Eq. (4.132).

Questions

4.1 Calculate the solution to $\ddot{x}+2\dot{x}+2x=\delta(t-\pi)$ with initial conditions $x(0)=1$, $\dot{x}(0)=0$.

4.2 Calculate the unit impulse response function for a critically damped system.

4.3 Calculate the unit impulse response of an overdamped system.

4.4 Calculate the response of $3\ddot{x}+c\dot{x}+12x=5\delta(t)$.

(1) for zero initial conditions when $c=16$;

(2) subject to the initial conditions $x(0)=0.01$ m and $v(0)=0$ when $c=12$.

4.5 Calculate the response of the system $3\ddot{x}+4\dot{x}+12x=2\delta(t)+4\delta(t-2)$ subject to the zero initial conditions.

4.6 Assume that a SDOF system is initially at rest in equilibrium position, find the response

(1) when a delayed step excitation $F(t)=\begin{cases}0 & t<t_0\\ F_0 & t\geqslant t_0\end{cases}$ is applied;

(2) when a delayed sine excitation $F(t)=\begin{cases}0 & t<t_0\\ F_0\sin[\omega(t-t_0)] & t\geqslant t_0\end{cases}$ is applied.

4.7 In the vibration testing of a structure, an impact hammer with a load cell to measure the impact force is used to cause excitation, as shown in Fig. 4.21. Assuming $m=5$ kg, $k=2\,000$ N/m, $c=10$ Ns/m and $F(t)=20\delta(t)$ N. Find the response of the system.

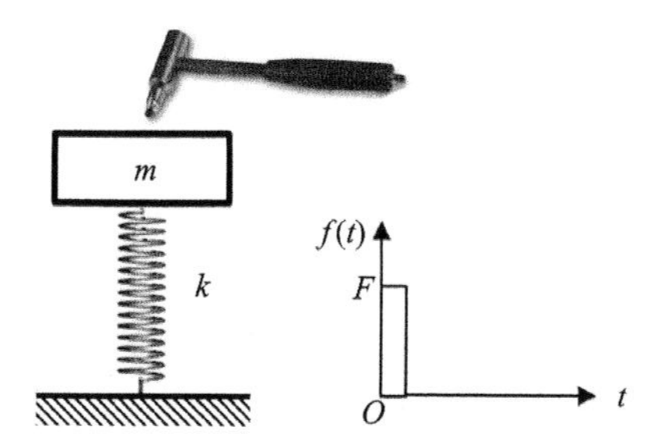

Fig. 4.21 Structural testing using an impact hammer

4.8 In Question 4.7, if the impact force is changed as $F(t)=20\delta(t)+10\,\delta(t-0.2)$ N, find the response of the structure.

4.9 Calculate the solution to $\ddot{x}+2\dot{x}+3x=\sin t+\delta(t-\pi)$ subject to the initial conditions $x(0)=0$ and $v(0)=1$.

4.10 Assume that a SDOF system is initially at rest in equilibrium position, find the response when a force $F(t)=F_0\exp(-at)$ is applied.

4.11 Consider a simple model of an airplane wing given in Fig. 4.22. The wing is approximated as vibrating back and forth in its plane, massless compared to the missile carriage system (of mass m). The modulus and the moment of inertia of the wing are approximated by E and I, respectively, and l is the length of the wing. The wing is modeled as a simple cantilever for the purpose of estimating the vibration resulting from the release of the missile, which is approximated by the impulse function $F\delta(t)$. Calculate the response for the case of an aluminum wing 2 m long with $m = 1\ 000$ kg, $\zeta = 0.01$, and $I = 0.5m^4$. Model F as 1 000 N lasting over 10^{-2} s. Modeling of wing vibration resulting from the release of a missile.

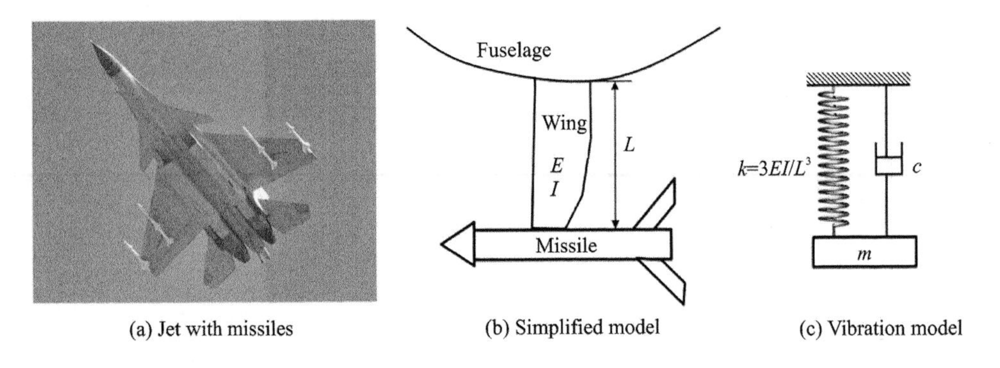

(a) Jet with missiles (b) Simplified model (c) Vibration model

Fig. 4.22 A simple model of an airplane wing

Chapter 5 Vibration of Multiple-Degree-Of-Freedom (MDOF) Systems

The simple SDOF systems analyzed in the Chapters 2—4 are very helpful to understand the general dynamic characteristics of vibrating systems. However, the SDOF systems are too simple to simulate most real systems, because the real systems always have more than just one degree of freedom. Most vibration structures and systems have many degrees of freedom. Therefore, more complex models are required to model their vibration behaviors.

By connecting several lumped-parameter models together, the simple SDOF systems can be constructed into a more complex Multiple-Degree-Of-Freedom (MDOF) system. Depending on the complexity of the model, the correct mathematical models can then be developed through the application of Newton's laws. Newton's laws are best used for models comprised of a few lumped parameter models described by a mass, spring, and damping values.

The motion of systems with many degrees of freedom, or nonlinear systems, always cannot be described by using simple formulas. Even when they can, the formulas are so long and complicated that you need a computer to evaluate them. For this reason, some MATLAB programs are presented in previous chapters. In this chapter, we also will use MATLAB to analyze the motion of these complex MDOF systems. You will find that it is actually quite straightforward. In fact, it is often easier than using the nasty formulas we derived for SDOF systems.

5.1 Free Vibration of Structures with Two-Degree-of-Freedom

First, a two-degree-of-freedom vibration system will be considered. This is because the addition of more degrees of freedom increases the labor of the solution procedure but does not introduce any new analytical principles. Initially, we will obtain the equations of motion for a two-degree-of-freedom system, then the natural frequencies and corresponding mode shapes will be presented. Some examples of two-degree-of-freedom models of vibrating structures are shown in Fig. 5.1.

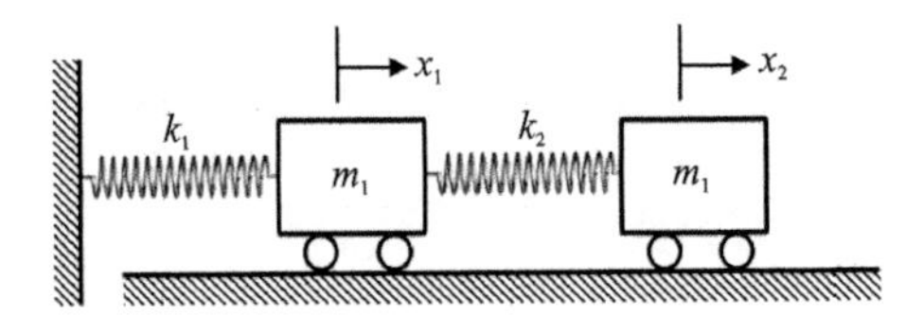

(a) Linear undamped system with coordinates x_1 and x_2

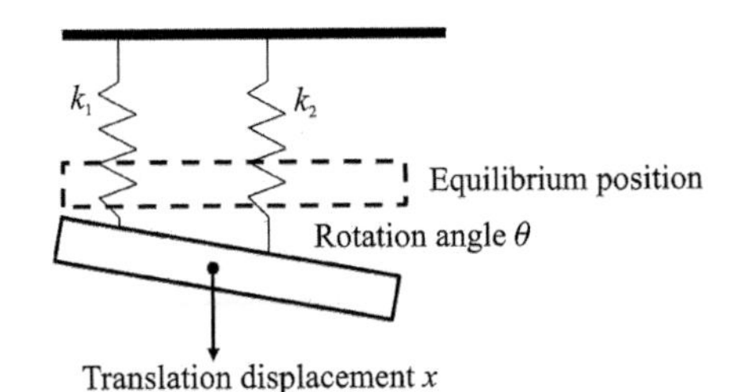

(b) System with combined translation and rotation with coordinates x and θ

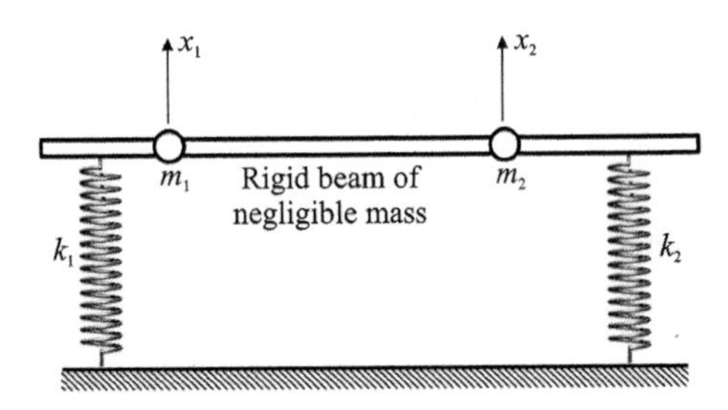

(c) Two degrees of freedom model with coordinates x_1 and x_2

Fig. 5.1 Some examples of two-degree-of-freedom models of vibrating structures

5.1.1 Equations of motion for free vibration of two-degree-of-freedom

The basic type of response of MDOF systems is free undamped vibration. Analogous to SDOF systems, the analysis of free vibration yields the natural frequencies of the system. We will consider systems without damping initially for following reasons:

(1) The math is easier.

(2) In practice the damping is often small.

(3) The main results are not too dependent on damping.

(4) Damping may be considered later.

Of the examples of two degrees of freedom models shown in Fig. 5.1(a)—(c), consider the system shown in Fig. 5.1(a). The free body diagrams are as shown in Fig. 5.2.

From Fig. 5.2, we can find that the equations of motion can be expressed as

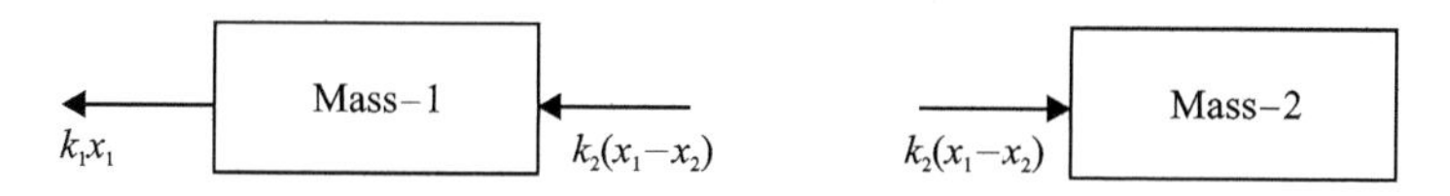

Fig. 5.2 Free body diagram for Mass-1 and Mass-2

follows:

$$m_1\ddot{x}_1 = -k_1x_1 - k_2(x_1 - x_2) \quad \text{for Mass-1} \tag{5.1}$$

$$m_2\ddot{x}_2 = k_2(x_1 - x_2) \quad \text{for Mass-2} \tag{5.2}$$

Eqs. (5.1) and (5.2) can be rewritten in the following matrix form:

$$\begin{bmatrix} m_1 & 0 \\ 0 & m_2 \end{bmatrix}\begin{bmatrix} \ddot{x}_1 \\ \ddot{x}_2 \end{bmatrix} + \begin{bmatrix} k_1+k_2 & -k_2 \\ -k_2 & k_2 \end{bmatrix}\begin{bmatrix} x_1 \\ x_2 \end{bmatrix} = \begin{bmatrix} 0 \\ 0 \end{bmatrix} \tag{5.3}$$

5.1.2 Free vibration analysis

When excited by an arbitrary initial state, it is interested to know whether m_1 and m_2 in Fig. 5.1(a) and Eq. (5.3) can oscillate harmonically with the same frequency and phase angle but with different amplitudes. Assuming that it is possible to have harmonic motion of m_1 and m_2 at the same frequency ω and the same phase angle ϕ. The solution of Eq. (5.3) can be expressed as follows:

$$x_1 = X_1\sin(\omega t + \phi) \tag{5.4a}$$

$$x_2 = X_2\sin(\omega t + \phi) \tag{5.4b}$$

Substituting Eq. (5.4) into Eq. (5.3), we get

$$\begin{bmatrix} m_1 & 0 \\ 0 & m_2 \end{bmatrix}\begin{bmatrix} -\omega^2X_1 \\ -\omega^2X_2 \end{bmatrix}\sin(\omega t + \phi) + \begin{bmatrix} k_1+k_2 & -k_2 \\ -k_2 & k_2 \end{bmatrix}\begin{bmatrix} X_1 \\ X_2 \end{bmatrix}\sin(\omega t + \phi)$$

$$= \left(\begin{bmatrix} -\omega^2m_1 & 0 \\ 0 & -\omega^2m_2 \end{bmatrix} + \begin{bmatrix} k_1+k_2 & -k_2 \\ -k_2 & k_2 \end{bmatrix}\right)\begin{bmatrix} X_1 \\ X_2 \end{bmatrix}$$

$$= \begin{bmatrix} -\omega^2m_1+k_1+k_2 & -k_2 \\ -k_2 & -\omega^2m_2+k_2 \end{bmatrix}\begin{bmatrix} X_1 \\ X_2 \end{bmatrix} = \begin{bmatrix} 0 \\ 0 \end{bmatrix} \tag{5.5}$$

From Eq. (5.5), we can find that it is satisfied by the trivial solution $X_1 = X_2 = 0$, which implies that there is no vibration. For a nontrivial solution of X_1 and X_2, the determinant of the coefficients matrix of X_1 and X_2 must be zero. So we get

$$\det\begin{bmatrix} -\omega^2m_1+k_1+k_2 & -k_2 \\ -k_2 & -\omega^2m_2+k_2 \end{bmatrix} = 0 \tag{5.6}$$

From Eq. (5.6), we are easy to obtain

$$(-\omega^2m_1+k_1+k_2)(-\omega^2m_2+k_2) - k_2^2 = 0 \tag{5.7}$$

Eq. (5.6) or Eq. (5.7) is called the frequency or characteristic equation because its solution yields the frequencies or the characteristic values of the system. Eq. (5.7) is a quadratic equation in ω^2 and thus gives two frequencies at which sinusoidal and nondecaying motion may occur without being forced. That is, the solutions of Eq. (5.7) give the natural frequencies two natural frequencies ω_{n1} and ω_{n2} for the system shown in Fig. 5.1(a).

Consider the case when $k_1 = k_2 = k$, and $m_1 = m_2 = m$. The frequency equation is simplified as

$$m\omega^4 - 3km\omega^2 + k^2 = 0 \tag{5.8}$$

The roots of Eq. (5.8) are given by

$$\omega_1^2 = \frac{3-\sqrt{5}}{2}\frac{k}{m} = 0.382\frac{k}{m} \tag{5.9a}$$

$$\omega_2^2 = \frac{3+\sqrt{5}}{2}\frac{k}{m} = 2.618\frac{k}{m} \tag{5.9b}$$

From Eq. (5.9), we can find that the system have a nontrivial harmonic solution for system in Eq. (5.3). The solutions ω_1 and ω_2 given in Eq. (5.9) are called as the natural frequencies of the system. Since natural frequencies are positive, by taking the positive square root in Eq. (5.9), we get $\omega_1 = \sqrt{0.382\frac{k}{m}}$ rad/s and $\omega_2 = \sqrt{2.618\frac{k}{m}}$ rad/s, respectively.

The values of X_1 and X_2 in Eq. (5.4) remain to be determined. These values depend on the natural frequencies ω_1 and ω_2. We shall denote the values of X_1 and X_2 corresponding to ω_1 as $X_1^{(1)}$ and $X_2^{(1)}$, and those corresponding values to ω_2 as $X_1^{(2)}$ and $X_2^{(2)}$. Furthermore, since Eq. (5.5) is homogenous, only the ratios $r_1 = X_2^{(1)}/X_1^{(1)}$ and $r_2 = X_2^{(2)}/X_2^{(2)}$ can be found. Substituting Eqs. (5.9a) and (5.9b) into Eq. (5.5), we get

$$\begin{aligned} r_1 &= \frac{X_2^{(1)}}{X_1^{(1)}} = \frac{-m\omega_1^2 + 2k}{k} = \frac{1+\sqrt{5}}{2} = 1.618 \\ r_2 &= \frac{X_2^{(2)}}{X_1^{(2)}} = \frac{-m\omega_2^2 + 2k}{k} = \frac{1-\sqrt{5}}{2} = -0.618 \end{aligned} \tag{5.10}$$

From Eq. (5.10), we get

$$\boldsymbol{X}^{(1)} = \begin{bmatrix} X_1 \\ X_2 \end{bmatrix}^{(1)} = \begin{bmatrix} 1 \\ r_1 \end{bmatrix} X_1^{(1)} \quad \text{for} \quad \omega_1^2 = 0.382\frac{k}{m} \tag{5.11a}$$

$$\boldsymbol{X}^{(2)} = \begin{bmatrix} X_1 \\ X_2 \end{bmatrix}^{(2)} = \begin{bmatrix} 1 \\ r_2 \end{bmatrix} X_1^{(2)} \quad \text{for} \quad \omega_2^2 = 2.618\frac{k}{m} \tag{5.11b}$$

$\boldsymbol{X}^{(1)}$ and $\boldsymbol{X}^{(2)}$ in Eq. (5.11) are termed as mode shapes (or eigenvectors) corresponding to the natural frequencies ω_1 and ω_2. Thus, the first mode of free vibration occurs at a frequency

$$\omega_1 = \sqrt{0.382 \frac{k}{m}} = 0.618\sqrt{\frac{k}{m}} \text{ rad/s} \quad \text{and} \quad X_2^{(1)} / X_1^{(1)} = 1.618$$

That is, the bodies move in phase with each other and with the amplitude ratio equal to 1.618 (as shown in Fig. 5.3).

The second mode of free vibration occurs at a frequency

$$\omega_2 = \sqrt{2.618 \frac{k}{m}} = 1.618\sqrt{\frac{k}{m}} \text{ rad/s} \quad \text{and} \quad X_2^{(1)} / X_1^{(1)} = -0.618$$

That is, the bodies move out of phase with each other, but with the amplitude ratio equal to 0.618. It should also be noticed that the natural frequencies are NOT dependent on the amplitude of vibration, and the ratio of amplitudes (the mode shape) is fixed for a given natural frequency.

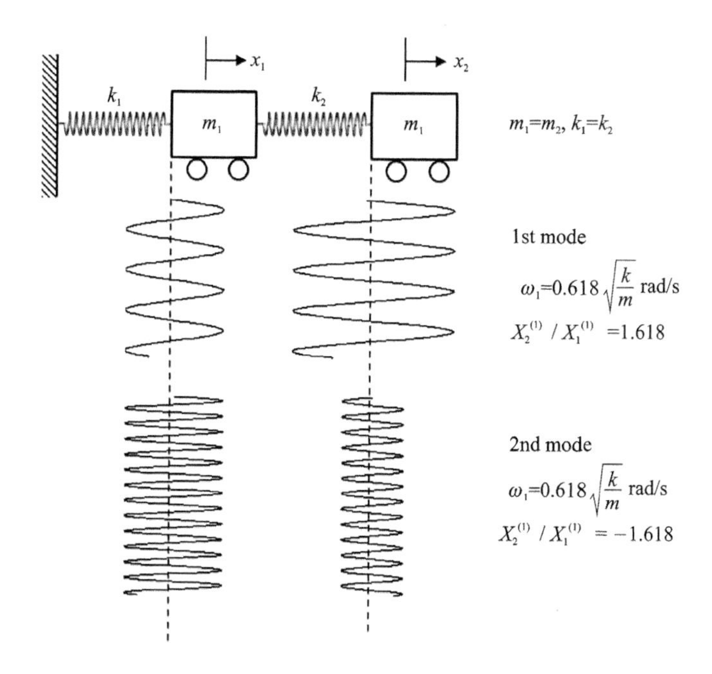

Fig. 5.3 Natural frequencies and mode shapes for two degrees of freedom vibration system

The mode shapes are shown in Fig. 5.4. In Fig. 5.4, the axial motion amplitude at each position is shown perpendicular to the axis of vibration for clarity of visualization. It should be noticed that Eq. (5.10) gives only the ratio of the amplitudes, and their absolute values are arbitrary if the ratio keeps the same.

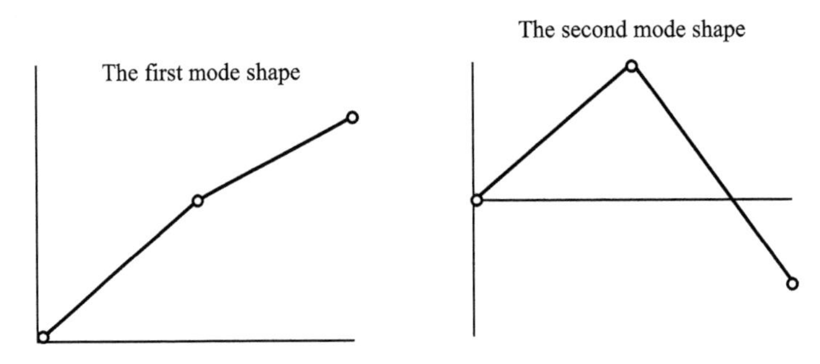

Fig. 5.4 The mode shapes

5.1.3 Free vibration responses

By using Eqs. (5.4) and (5.11), the free vibration solution (the motion in time) can be expressed as follows:

$$\boldsymbol{x}^{(1)} = \begin{bmatrix} x_1^{(1)} \\ x_2^{(1)} \end{bmatrix} = \begin{bmatrix} 1 \\ r_1 \end{bmatrix} X_1^{(1)} \sin(\omega_1 t + \phi_1) \tag{5.12a}$$

$$\boldsymbol{x}^{(2)} = \begin{bmatrix} x_1^{(2)} \\ x_2^{(2)} \end{bmatrix} = \begin{bmatrix} 1 \\ r_2 \end{bmatrix} X_1^{(2)} \sin(\omega_2 t + \phi_2) \tag{5.12b}$$

where $X_1^{(1)}$, $X_2^{(1)}$, ϕ_1 and ϕ_2 can be determined by the initial conditions.

The equation of motion in Eq. (5.3) involves the second-order time derivatives. So we need to specify two initial conditions for each mass. For general initial conditions, both modes will be excited. The resulting motion, which is given by the general solution of Eq. (5.3) can be obtained by a linear superposition of the two normal modes in Eq. (5.12).

$$\boldsymbol{x} = c_1 \boldsymbol{x}^{(1)} + c_2 \boldsymbol{x}^{(2)} \tag{5.13}$$

where c_1 and c_2 are the real constants. Since Eq. (5.13) already involves the unknown constants $X_1^{(1)}$ and $X_2^{(1)}$ (see Eq. (5.12)), we can choose $c_1 = c_2 = 1$ without loss of generality. Thus the components of the vector $\boldsymbol{x}$ can be expressed as

$$\boldsymbol{x} = \boldsymbol{x}^{(1)} + \boldsymbol{x}^{(2)} = \begin{bmatrix} 1 \\ r_1 \end{bmatrix} X_1^{(1)} \sin(\omega_1 t + \phi_1) + \begin{bmatrix} 1 \\ r_2 \end{bmatrix} X_1^{(2)} \sin(\omega_2 t + \phi_2) \tag{5.14}$$

For example, for the system considered above, if one body is displaced a distance X_0 and released, so the initial conditions can be written as

$$\boldsymbol{x}(t=0) = \begin{bmatrix} x_1(0) \\ x_2(0) \end{bmatrix} = \begin{bmatrix} X_0 \\ 0 \end{bmatrix} \quad \text{and} \quad \boldsymbol{v}(t=0) = \begin{bmatrix} \dot{x}_1(0) \\ \dot{x}_2(0) \end{bmatrix} = \begin{bmatrix} 0 \\ 0 \end{bmatrix} \tag{5.15}$$

Remembering that in this system, $\omega_1 = 0.618\sqrt{\frac{k}{m}}$ rad/s, $\omega_2 = 1.618\sqrt{\frac{k}{m}}$ rad/s, $r_1 = 1.618$ and $r_2 = -0.618$ (see Eqs. (5.9) and (5.10)). From Eqs. (5.14) and

(5.15), we get

$$\begin{bmatrix} x_1(0) \\ x_2(0) \end{bmatrix} = \begin{bmatrix} X_0 \\ 0 \end{bmatrix} = \begin{bmatrix} 1 \\ 1.618 \end{bmatrix} X_1^{(1)} \sin \phi_1 + \begin{bmatrix} 1 \\ -0.618 \end{bmatrix} X_1^{(2)} \sin \phi_2 \tag{5.16a}$$

$$\begin{bmatrix} \dot{x}_1(0) \\ \dot{x}_2(0) \end{bmatrix} = \begin{bmatrix} 0 \\ 0 \end{bmatrix} = \begin{bmatrix} 1 \\ 1.618 \end{bmatrix} X_1^{(1)} [\omega_1 \cos \phi_1] + \begin{bmatrix} 1 \\ -0.618 \end{bmatrix} X_1^{(2)} [\omega_2 \cos \phi_2] \tag{5.16b}$$

From Eq. (5.16), it is easy to find that

$$X_1^{(1)} = 0.276X_0, \quad X_1^{(2)} = 0.724X_0, \quad \phi_1 = \phi_2 = \frac{\pi}{2} \tag{5.17}$$

Substituting Eq. (5.17) into Eq. (5.14), we get

$$\boldsymbol{x}(t) = \begin{bmatrix} 0.276X_0\cos(\omega_1 t) + 0.724X_0\cos(\omega_2 t) \\ 0.276r_1X_0\cos(\omega_1 t) + 0.724r_2X_0\cos(\omega_2 t) \end{bmatrix} \tag{5.18}$$

Clearly, for general initial conditions, both natural frequencies are excited and the motion of each mass has two harmonic components. Fig. 5.5 shows the time responses of the system for $k = m = X_0 = 1$.

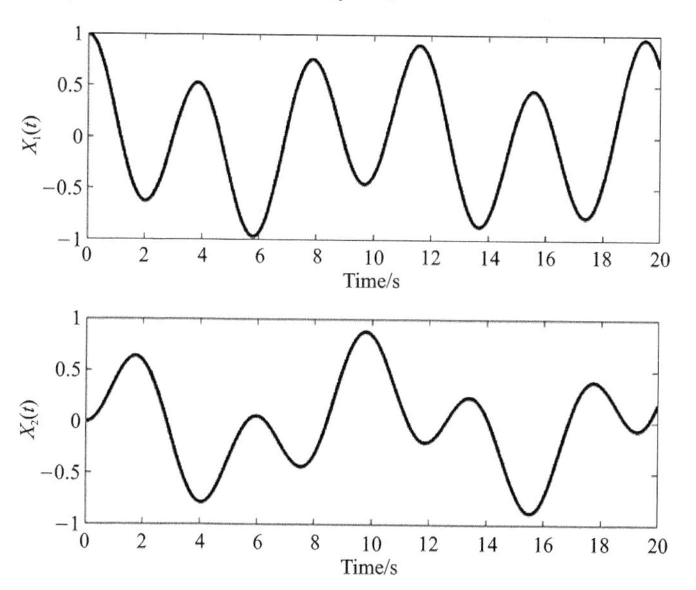

Fig. 5.5 The time response for Eq. (5.18)

5.2 Modelling of a System with n Degrees of Freedom

5.2.1 Governing equations

A system with n degrees of freedom is governed by n second (s) order differential equations. The Newton's second law can also be used to derive the mathematical model of such a system. Consider a three-degree-of-freedom system shown in Fig. 5.6, assume the masses are constrained to move along the vertical direction. The three external forces $f_1(t)$, $f_2(t)$ and $f_3(t)$ are acting on m_1, m_2 and m_3, respectively. The variables x_1, x_2 and x_3 are the displacements of m_1, m_2 and m_3 from their static equilibrium positions. Similar to the SDOF systems, the MDOF systems are also advisable expressed the motion of the systems from their static equilibrium position.

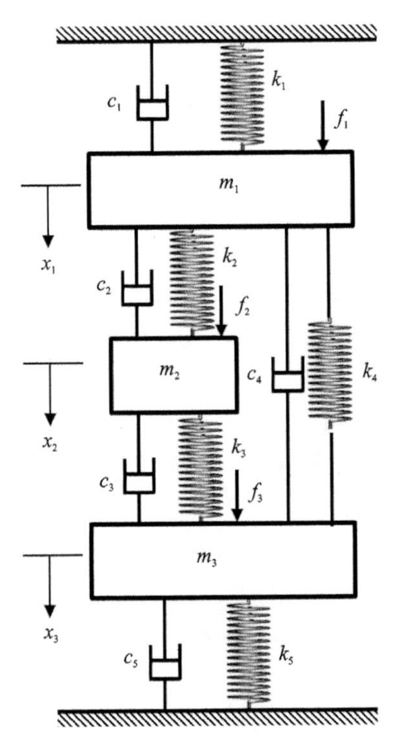

Fig. 5.6 A three-degree-of-freedom system

Then the free body diagram for the mass m_1 is shown in Fig. 5.7.

Applying the Newton's second law to mass m_1, we have

$$m_1\ddot{x}_1 = -k_1x_1 - c_1\dot{x}_1 + k_2(x_2 - x_1) + c_2(\dot{x}_2 - \dot{x}_1) + k_4(x_3 - x_1) + c_4(\dot{x}_3 - \dot{x}_1) + f_1(t) \tag{5.19}$$

To rewrite Eq. (5.19), we get

$$m_1\ddot{x}_1 + (c_1 + c_2 + c_4)\dot{x}_1 - c_2\dot{x}_2 - c_4\dot{x}_3 + (k_1 + k_2 + k_4)x_1 - k_2x_2 - k_4x_3 = f_1(t) \tag{5.20}$$

Similarly, the free body diagram for m_2 is shown in Fig. 5.8. From Fig. 5.8, we can find that

$$m_2\ddot{x}_2 - c_2\dot{x}_1 + (c_2 + c_3)\dot{x}_2 - c_3\dot{x}_3 - k_2x_1 + (k_2 + k_3)x_2 - k_3x_3 = f_2(t) \tag{5.21}$$

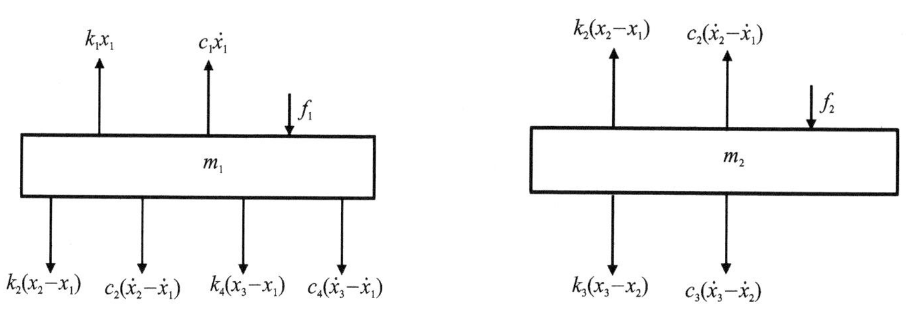

Fig. 5.7 The free body diagram for the mass m_1 **Fig. 5.8 The free body diagram for the mass m_2**

Similar as above, the free body diagram for m_3 is shown in Fig. 5.9.

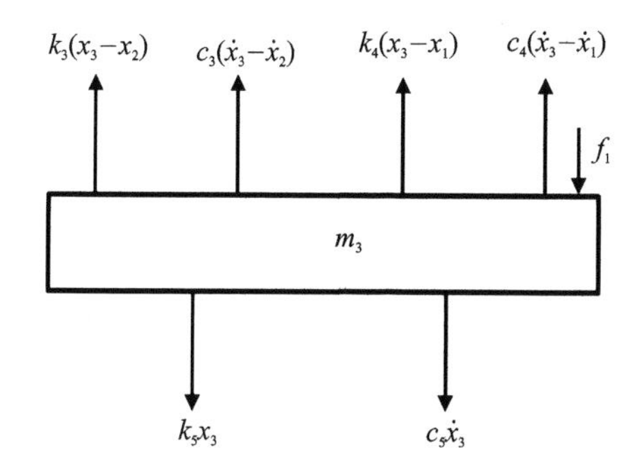

Fig. 5.9 The free body diagram for the mass m_3

From Fig. 5.9, we are easy to find that

$$m_3\ddot{x}_3 - c_4\dot{x}_1 - c_3\dot{x}_2 + (c_3 + c_4 + c_5)\dot{x}_3 - k_4x_1 - k_3x_2 + (k_3 + k_4 + k_5)x_3 = f_3(t) \tag{5.22}$$

Eqs. (5.20)—(5.22) can be written in the following matrix form:

$$\begin{bmatrix} m_1 & 0 & 0 \\ 0 & m_2 & 0 \\ 0 & 0 & m_3 \end{bmatrix} \begin{bmatrix} \ddot{x}_1(t) \\ \ddot{x}_2(t) \\ \ddot{x}_3(t) \end{bmatrix} + \begin{bmatrix} c_1 + c_2 + c_4 & -c_2 & -c_4 \\ -c_2 & c_2 + c_3 & -c_3 \\ -c_4 & -c_3 & c_3 + c_4 + c_5 \end{bmatrix} \begin{bmatrix} \dot{x}_1(t) \\ \dot{x}_2(t) \\ \dot{x}_3(t) \end{bmatrix} +$$
$$\begin{bmatrix} k_1 + k_2 + k_4 & -k_2 & -k_4 \\ -k_2 & k_2 + k_3 & -k_3 \\ -k_4 & -k_3 & k_3 + k_4 + k_5 \end{bmatrix} \begin{bmatrix} x_1(t) \\ x_2(t) \\ x_3(t) \end{bmatrix} = \begin{bmatrix} f_1(t) \\ f_2(t) \\ f_3(t) \end{bmatrix} \tag{5.23}$$

Assume that

$$\boldsymbol{M} = \begin{bmatrix} m_1 & 0 & 0 \\ 0 & m_2 & 0 \\ 0 & 0 & m_3 \end{bmatrix} \tag{5.24a}$$

$$\boldsymbol{C}=\begin{bmatrix} c_1+c_2+c_4 & -c_2 & -c_4 \\ -c_2 & c_2+c_3 & -c_3 \\ -c_4 & -c_3 & c_3+c_4+c_5 \end{bmatrix} \tag{5.24b}$$

$$\boldsymbol{K}=\begin{bmatrix} k_1+k_2+k_4 & -k_2 & -k_4 \\ -k_2 & k_2+k_3 & -k_3 \\ -k_4 & -k_3 & k_3+k_4+k_5 \end{bmatrix} \tag{5.24c}$$

$$\boldsymbol{x}=\begin{bmatrix} \ddot{x}_1(t) \\ \ddot{x}_2(t) \\ \ddot{x}_3(t) \end{bmatrix} \tag{5.24d}$$

$$\boldsymbol{f}=\begin{bmatrix} f_1(t) \\ f_2(t) \\ f_3(t) \end{bmatrix} \tag{5.24e}$$

and then Eq. (5.23) can be symbolically written in the following compact matrix form:

$$\boldsymbol{M}\ddot{\boldsymbol{x}}+\boldsymbol{C}\dot{\boldsymbol{x}}+\boldsymbol{K}\boldsymbol{x}=\boldsymbol{f} \tag{5.25}$$

Three matrices $\boldsymbol{M}$, $\boldsymbol{C}$ and $\boldsymbol{K}$ are called as mass matrix, damping matrix and stiffness matrix, respectively. For the three-degree-of-freedom system analyzed here. $\boldsymbol{x}$, $\boldsymbol{M}$, $\boldsymbol{C}$, $\boldsymbol{K}$ and $\boldsymbol{f}$ are given by Eq. (5.24). However, Eq. (5.25) holds for a general vibration system with many degrees of freedom.

Example 5.1

In this example, a three-degree-of-freedom system is shown in Fig. 5.10. The notation m_i represents mass, c_i represents a damper, and k_i represents a spring. The forces related to the dampers are represented by F_{ci} and the forces related to the springs are represented by F_{si}. Find the mathematical model for a multiple-degree-of-freedom model by using Newton's laws.

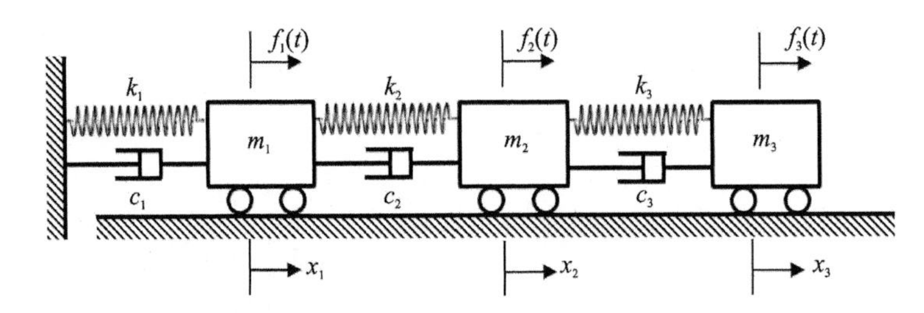

Fig. 5.10　A multiple-degree-of-freedom lumped parameter model

Solution:

The first step in setting up this solution is to break the model into a free body diagram for each mass, as shown in Fig. 5.11.

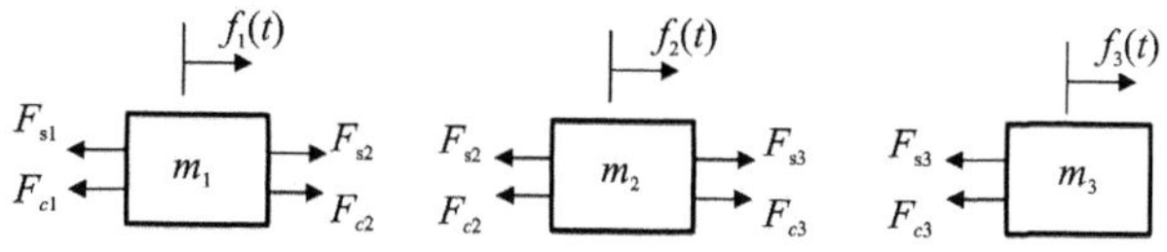

Fig. 5.11 A multiple-degree-of-freedom lumped parameter model free-body diagrams

After each free body diagram has been drawn and annotated, the equation of motion for each particle must be defined by Newton's second law.

For m_1

$$m_1\ddot{x}_1+(c_1+c_2)\dot{x}_1-c_2\dot{x}_2-c_5\dot{x}_3+(k_1+k_2)x_1-k_2x_2=f_1(t) \quad (5.26)$$

For m_2

$$m_2\ddot{x}_2-c_2\dot{x}_1+(c_2+c_3)\dot{x}_2-c_3\dot{x}_3-k_2x_1+(k_2+k_3)x_2-k_3x_3=f_2(t) \quad (5.27)$$

For m_3

$$m_3\ddot{x}_3-c_3\dot{x}_2+c_3\dot{x}_3-k_3x_2+k_3x_3=f_3(t) \quad (5.28)$$

Eqs. (5.26)—(5.28) can be written in the following matrix form:

$$\begin{bmatrix} m_1 & 0 & 0 \\ 0 & m_2 & 0 \\ 0 & 0 & m_3 \end{bmatrix}\begin{bmatrix} \ddot{x}_1(t) \\ \ddot{x}_2(t) \\ \ddot{x}_3(t) \end{bmatrix}+\begin{bmatrix} c_1+c_2 & -c_2 & 0 \\ -c_2 & c_2+c_3 & -c_3 \\ 0 & -c_3 & c_3 \end{bmatrix}\begin{bmatrix} \dot{x}_1(t) \\ \dot{x}_2(t) \\ \dot{x}_3(t) \end{bmatrix}+$$

$$\begin{bmatrix} k_1+k_2 & -k_2 & 0 \\ -k_2 & k_2+k_3 & -k_3 \\ 0 & -k_3 & k_3 \end{bmatrix}\begin{bmatrix} x_1(t) \\ x_2(t) \\ x_3(t) \end{bmatrix}=\begin{bmatrix} f_1(t) \\ f_2(t) \\ f_3(t) \end{bmatrix} \quad (5.29)$$

The mathematical model for the MDOF system can be defined through the use of matrix notation. This will allow for the evaluation and determination of the exact solution. These matrices define the mass $\boldsymbol{M}$, stiffness $\boldsymbol{K}$, damping $\boldsymbol{C}$. The corresponding excitation vector is defined as $\boldsymbol{f}$.

The mass matrix $\boldsymbol{M}$ is

$$\boldsymbol{M}=\begin{bmatrix} m_1 & 0 & 0 \\ 0 & m_2 & 0 \\ 0 & 0 & m_3 \end{bmatrix} \quad (5.30a)$$

The damping matrix $\boldsymbol{C}$ is

$$\boldsymbol{C}=\begin{bmatrix} c_1+c_2 & -c_2 & 0 \\ -c_2 & c_2+c_3 & -c_3 \\ 0 & -c_3 & c_3 \end{bmatrix} \quad (5.30b)$$

The stiffness matrix $\boldsymbol{K}$ is

$$\mathbf{K}=\begin{bmatrix} k_1+k_2 & -k_2 & 0 \\ -k_2 & k_2+k_3 & -k_3 \\ 0 & -k_3 & k_3 \end{bmatrix} \tag{5.30c}$$

The excitation vector $\boldsymbol{f}$ is

$$\boldsymbol{f}=\begin{bmatrix} f_1(t) \\ f_2(t) \\ f_3(t) \end{bmatrix} \tag{5.30d}$$

5.2.2 Coordinate coupling

It should be noticed that the motion equation is not unique for a given MDOF system. It depends on the choosing of the coordinates. Consider a double pendulum shown in Fig. 5.12.

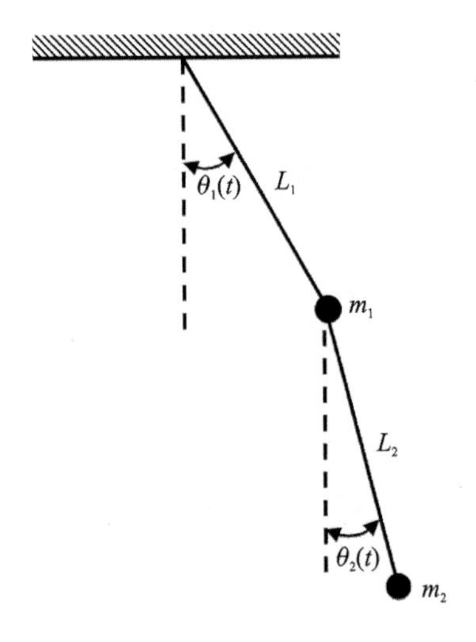

Fig. 5.12 A double pendulum with coordinates θ_1 and θ_2

First, assume that the double pendulum is with coordinates θ_1 and θ_2, and the angle displacements θ_1 and θ_2 are small (such as $\sin\theta_1=\theta_1$ and $\sin\theta_2=\theta_2$). The free body diagram for m_1 is shown in Fig. 5.13. Clearly, the tension force T_2 is

$$T_2=m_2 g \tag{5.31}$$

If we apply the Newton's second law in the direction perpendicular to T_1 (see Fig. 5.13), the component of the tension force T_1 vanishes in the direction, and we have

$$m_1(L_1\ddot{\theta}_1)=-m_1 g\theta_1+m_2 g(\theta_2-\theta_1) \tag{5.32}$$

The free body diagram for m_2 is shown in Fig. 5.14.

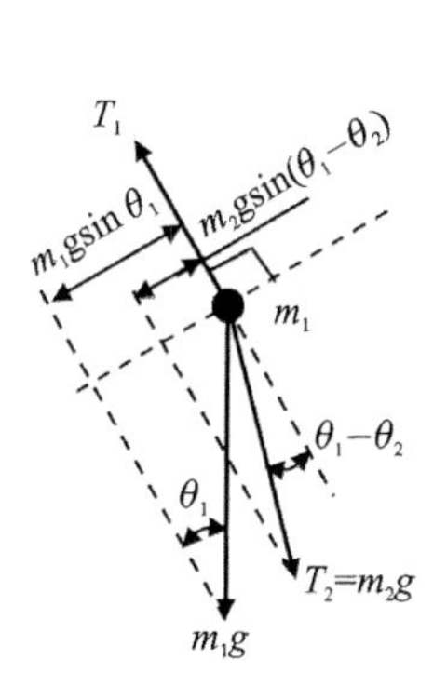

Fig. 5.13 The free body diagram for m_1

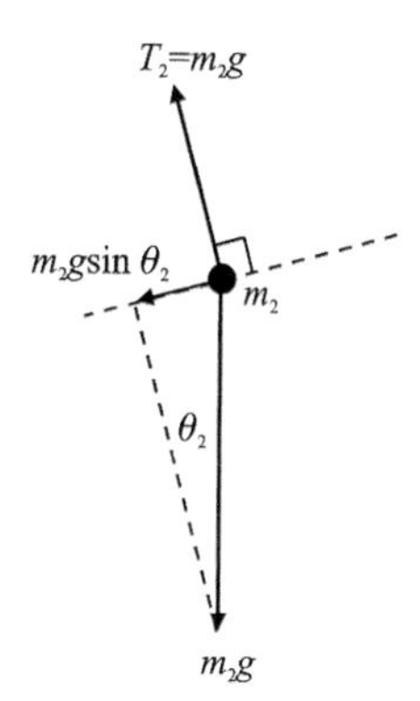

Fig. 5.14 The free body diagram for m_2

If we apply the Newton's second law in the direction perpendicular to T_2 (see Fig. 5.14), the component of the tension force T_2 vanishes in the direction, and we have

$$m_2(L_1\ddot{\theta}_1+L_2\ddot{\theta}_2)=-m_2g\theta_2 \tag{5.33}$$

The differential equations of motionin Eqs. (5.32) and (5.33) can be written in matrix form as follows:

$$\begin{bmatrix} m_1L_1 & 0 \\ L_1 & L_2 \end{bmatrix}\begin{bmatrix} \ddot{\theta}_1 \\ \ddot{\theta}_2 \end{bmatrix}+\begin{bmatrix} (m_1+m_2)g & -m_2g \\ 0 & g \end{bmatrix}\begin{bmatrix} \theta_1 \\ \theta_2 \end{bmatrix}=\begin{bmatrix} 0 \\ 0 \end{bmatrix} \tag{5.34}$$

Eq. (5.34) can be rewritten as matrix form:

$$\boldsymbol{M}\ddot{\boldsymbol{\theta}}+\boldsymbol{K}\boldsymbol{\theta}=\boldsymbol{0} \tag{5.35}$$

where

$$\boldsymbol{M}=\begin{bmatrix} m_1L_1 & 0 \\ L_1 & L_2 \end{bmatrix},\ \boldsymbol{K}=\begin{bmatrix} (m_1+m_2)g & -m_2g \\ 0 & g \end{bmatrix} \quad \text{and} \quad \boldsymbol{\theta}=\begin{bmatrix} \theta_1 \\ \theta_2 \end{bmatrix}$$

We see that here neither $\boldsymbol{M}$ nor $\boldsymbol{K}$ are symmetric. However, it is possible to rewrite the $\boldsymbol{M}$ or $\boldsymbol{K}$ matrices into symmetric form. It can be done by replace $m_2g\theta_2$ in Eq. (5.32) by using Eq. (5.33), we have

$$m_1(L_1\ddot{\theta}_1)=-m_1g\theta_1-m_2g\theta_1-m_2(L_1\ddot{\theta}_1+L_2\ddot{\theta}_2) \tag{5.36a}$$

or

$$(m_1L_1+m_2L_1)\ddot{\theta}_1+L_2\ddot{\theta}_2+(m_1g+m_2g)\theta_1=0 \tag{5.36b}$$

The differential equations of motion in Eqs. (5.36) and (5.33) can also be written in matrix form as follows:

$$\underbrace{\begin{bmatrix} m_1L_1+m_2L_1 & L_2 \\ L_1 & L_1 \end{bmatrix}}_{\boldsymbol{M}\text{ matrix}}\begin{bmatrix} \ddot{\theta}_1 \\ \ddot{\theta}_2 \end{bmatrix}+\underbrace{\begin{bmatrix} (m_1+m_2)g & 0 \\ 0 & g \end{bmatrix}}_{\boldsymbol{K}\text{ matrix}}\begin{bmatrix} \theta_1 \\ \theta_2 \end{bmatrix}=\begin{bmatrix} 0 \\ 0 \end{bmatrix} \tag{5.37}$$

From Eq. (5. 37), we can find that the $\boldsymbol{K}$ matrix is symmetric and uncoupled now.

Another way toexpress the double pendulum motion is to use the coordinates x_1 and x_2, the displacements of m_1 and m_2 from the static equilibrium position, respectively, as shown in Fig. 5. 15.

Clearly, the relations between θ_i and $x_i (i = 1, 2)$ can be expressed as

$$\sin[\theta_1(t)] = \frac{x_1(t)}{L_1}, \quad \sin[\theta_2(t)] = \frac{x_2(t) - x_1(t)}{L_2} \tag{5.38a}$$

According to small vibration assumption, Eq. (5. 38a) can be simplified as

$$\theta_1(t) \simeq \sin[\theta_1(t)] = \frac{x_1(t)}{L_1}, \quad \theta_2(t) \simeq \sin[\theta_2(t)] = \frac{x_2(t) - x_1(t)}{L_2} \tag{5.38b}$$

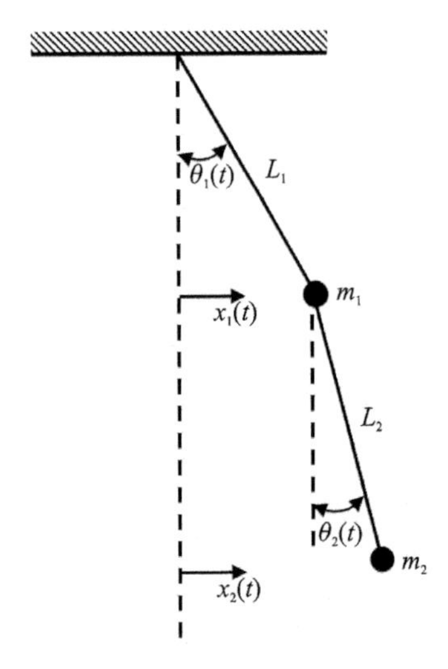

Fig. 5. 15 Thedouble pendulum with coordinates x_1 and x_2

Notice that the lengths L_1 and L_2 are constants, the angle accelerations can be obtained from Eq. (3. 38b), such as

$$\ddot{\theta}_1(t) = \frac{\ddot{x}_1(t)}{L_1}, \quad \ddot{\theta}_2(t) = \frac{\ddot{x}_2(t) - \ddot{x}_1(t)}{L_2} \tag{5.38c}$$

Substituting Eqs. (5. 38b) and (5. 38c) into Eqs. (5. 32) and (5. 33) respectively, we can find that the motion of m_1 and m_2 are governed by

$$m_1\ddot{x}_1 + \left(\frac{m_1 g}{L_1} + \frac{m_2 g}{L_1} + \frac{m_2 g}{L_2}\right)x_1 - \frac{m_2 g}{L_2}x_2 = 0 \tag{5.39a}$$

$$m_2\ddot{x}_2 - \frac{m_2 g}{L_2}x_1 + \frac{m_2 g}{L_2}x_2 = 0 \tag{5.39b}$$

Clearly, the matrix form of Eq. (5. 39) is

$$\begin{bmatrix} m_1 & 0 \\ 0 & m_2 \end{bmatrix}\begin{bmatrix} \ddot{x}_1 \\ \ddot{x}_2 \end{bmatrix}+\begin{bmatrix} \left(\dfrac{m_1}{L_1}+\dfrac{m_2}{L_1}+\dfrac{m_2}{L_2}\right)g & -\dfrac{m_2 g}{L_2} \\ -\dfrac{m_2 g}{L_2} & \dfrac{m_2 g}{L_2} \end{bmatrix}\begin{bmatrix} x_1 \\ x_2 \end{bmatrix}=\begin{bmatrix} 0 \\ 0 \end{bmatrix} \tag{5.40a}$$

Eq. (5.40a) can be rewritten as matrix form:

$$\boldsymbol{M}\ddot{\boldsymbol{x}}+\boldsymbol{K}\boldsymbol{x}=\boldsymbol{0} \tag{5.40b}$$

where

$$\boldsymbol{M}=\begin{bmatrix} m_1 & 0 \\ 0 & m_2 \end{bmatrix}, \quad \boldsymbol{K}=\begin{bmatrix} \left(\dfrac{m_1}{L_1}+\dfrac{m_2}{L_1}+\dfrac{m_2}{L_2}\right)g & -\dfrac{m_2 g}{L_2} \\ -\dfrac{m_2 g}{L_2} & \dfrac{m_2 g}{L_2} \end{bmatrix}$$

As expected, the system in Eq. (5.40) which expresses the motion of the double pendulum, measured from the static equilibrium position, has symmetric mass and stiffness matrices, and the mass matrix $\boldsymbol{M}$ is uncoupled. We have shown for the double pendulum that there exist three forms, namely Eqs. (5.35), (5.37) and (5.40), describing the motion of the same system. In fact, there are many valid forms which describe the motion of the same vibratory system. They are all related by scaling of the equations and transforming the degrees of freedom.

According to above analysis, the following characteristics of these systems can be found:

(1)The system vibrates in its own natural way regardless of the coordinates used. The choice of the coordinates is a mere convenience.

(2)From Eqs. (5.35), (5.37) and (5.40), it is clear that the nature of the coupling depends on the coordinates used and is not an inherent property of the system. It is possible to choose a system of coordinates which give equations of motion that are uncoupled both statically and dynamically. Such coordinates are called **principal** or **natural coordinates**, which will be discussed in Section 5.6.

Before moving on to the solution of MDOF systems, we will discuss how to obtain the mass and stiffness matrices by the influence coefficient method and the Lagrange's equation.

5.3 Influence Coefficient Method

The equations of motion of a MDOF system can also be obtained in terms of influence coefficients, which are extensively used in structural engineering. Basically, one set of influence coefficients can be associated with each of the matrices involved in the equations of motion. The influence coefficients associated with the stiffness and mass matrices are, respectively, known as the stiffness and inertia influence coefficients.

5.3.1 The stiffness influence coefficients

For a simple linear spring, the force necessary to cause a unit displacement is called the stiffness of the spring. In more complex systems, we can express the relation between the displacement at a point and the forces acting at various other points of the system by means of stiffness influence coefficients. The stiffness influence coefficient is defined as the force at point i due to a unit displacement at point j when all the points other than the point j are fixed. Using this definition, for the spring-mass system shown in Fig. 5.16, suppose the accelerations $\ddot{x}_j$ are zero at a particular instant, so that the inertia effects are absent. The stiffness matrix $\boldsymbol{K}$ is given under these circumstances, by the constitutive relation for the spring elements. So the total force at point i, can be found by summing up the forces due to all displacements x_j ($j=1, 2, \cdots, n$) as follows:

$$F_i = \sum_{j=1}^{n} k_{ij} x_j \quad \text{for} \quad i = 1, 2, \cdots, n \tag{5.41}$$

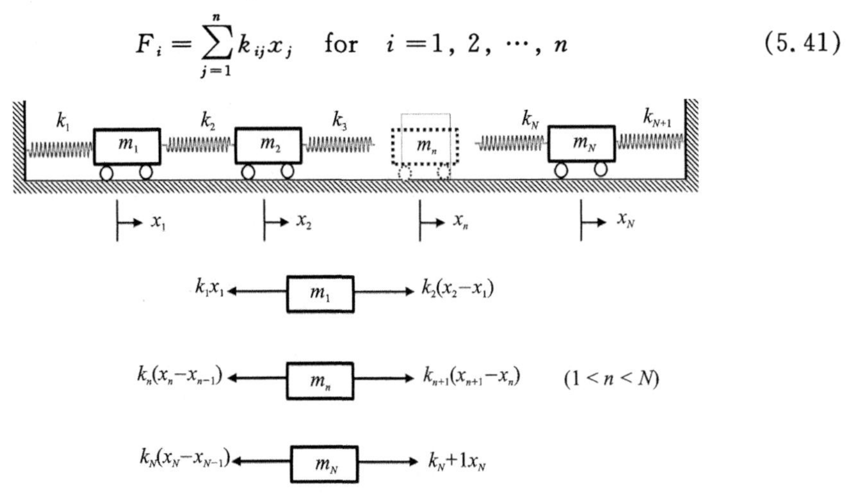

Fig. 5.16 A MDOF spring-mass system

Eq. (5.41) can be rewritten as matrix form:

$$\boldsymbol{F} = \boldsymbol{K}\boldsymbol{x} \tag{5.42}$$

where $\boldsymbol{F}$ is the displacement vector defined in Eq. (5.31), and $\boldsymbol{K}$ is the stiffness matrix given by

$$\boldsymbol{K} = \begin{bmatrix} k_{11} & k_{12} & \cdots & k_{1N} \\ k_{21} & k_{22} & \cdots & k_{2N} \\ \vdots & \vdots & & \vdots \\ k_{N1} & k_{N2} & \cdots & k_{NN} \end{bmatrix} \tag{5.43}$$

The following aspects of stiffness influence coefficients are to be noted:

(1) Since the force required at point i to cause a unit deflection at point j and zero deflection at all other points is the same as the force required at point j to cause a unit deflection at point i and zero deflection at all other points (reciprocity theorem), we have $k_{ij} = k_{ji}$.

(2) The stiffness influence coefficients can be calculated by applying the principles of statics and solid mechanics.

(3) The stiffness influence coefficients for torsional systems can be defined in terms of unit angular displacement and the torque that causes the angular displacement. For example, in a multi-rotor torsional system, it can be defined as the torque at point i (rotor i) due to a unit angular displacement at point j and zero angular displacement at all other points.

The stiffness influence coefficients of a MDOF system can be determined as follows:

(1) Assume the displacement $x_j = 1$ ($j = 1$ to start with) and $x_k = 0$ for all $k \neq j$.

(2) Determine the force vector $\boldsymbol{f}$ to maintain the system in the assumed configuration. By definition, $\boldsymbol{f}$ = the jth column of $\boldsymbol{K}$ matrix.

(3) After completing step (1) for $j = 1$, the procedure is repeated for $j = 2, 3, \cdots, n$.

In Fig. 5.16, we can find that $K_{n,n} = k_n + k_{n+1}$, $K_{n,n+1} = k_{n+1,n} = -k_n$, and the other elements in matrix $\boldsymbol{K}$ are zero.

Mass matrix, which is used in the case of translatory motions, can be generalized asinertia matrix $\boldsymbol{M}$ in order to include rotatory motions as well. To determine $\boldsymbol{M}$ for the systems shown in Fig. 5.16, suppose the deflections x_n ($n = 1, 2, \cdots, N$) are zero at a particular instant, so that the springs are in their static equilibrium configuration. Under these conditions, the equation of motion becomes

$$\boldsymbol{F} = \boldsymbol{M}\ddot{\boldsymbol{x}} \tag{5.44}$$

Similar to stiffness influence coefficients, the mass matrix of a MDOF system can be determined as follows:

(1) Assume the acceleration $\ddot{x}_j = 1$ ($j = 1$ to start with) and $\ddot{x}_k = 0$ for all $k \neq 1$.

(2) Determine the force vector $\boldsymbol{f}$ to maintain the system in the assumed configuration. By definition, $\boldsymbol{f}$ = the jth column of $\boldsymbol{M}$ matrix.

(3) After completing step (1) for $j = 1$, the procedure is repeated for $j = 2, 3, \cdots, n$.

By using above steps, it is easy to find the mass matrix for system in Fig. 5.16 is a diagonal matrix with diagonal elements $M_{n,n} = m_n$.

5.3.2 Flexibility influence coefficients

In some cases, it is more simple and convenient to rewrite the equations of motion using the inverse of the stiffness matrix (known as the flexibility matrix) or the inverse of the mass matrix. The influence coefficients corresponding to the inverse stiffness matrix are called the **flexibility influence coefficients**, and those corresponding to the inverse mass matrix are known as the inverse inertia coefficients.

Let the system be acted on by just one force F_j, and let the displacement at point i (i. e., mass m_i) due to F_j be x_{ij}. The flexibility influence coefficients, denoted as a_{ij}, is defined as the deflection at point i due to a unit load at point j. Clearly, we have

$$x_{ij} = a_{ij} F_j \tag{5.45}$$

If there are several forces F_j ($j = 1, 2, \cdots, n$) at different points of the system, the total deflection at any point i can be found by summing up the contributions of all forces F_j, that is

$$x_i = \sum_{j=1}^{n} a_{ij} F_j = \sum_{j=1}^{n} x_{ij} \tag{5.46}$$

Eq. (5.46) can be rewritten as matrix form:

$$\boldsymbol{x} = \boldsymbol{AF} \tag{5.47}$$

where $\boldsymbol{x}$ and $\boldsymbol{F}$ are the displacement and force vectors, and $\boldsymbol{A}$ is the flexibility matrix given by

$$\boldsymbol{A} = \begin{bmatrix} a_{11} & a_{12} & \cdots\cdots & a_{1n} \\ a_{21} & a_{22} & \cdots & a_{2n} \\ \vdots & \vdots & & \vdots \\ a_{n1} & a_{n2} & \cdots & a_{nn} \end{bmatrix} \tag{5.48}$$

Substituting Eq. (5.42) into Eq. (5.47), we get

$$\boldsymbol{AK} = \boldsymbol{I} \tag{5.49}$$

where $\boldsymbol{I}$ denotes the unit matrix.

From Eq. (5.49), we can find that

$$\boldsymbol{A} = \boldsymbol{K}^{-1} \tag{5.50}$$

From Eq. (5.50), we can find that the flexibility matrix $\boldsymbol{A}$ is the inverse of the stiffness matrix $\boldsymbol{K}$. Similar to before, the flexibility influence coefficients of a torsional system can also be defined in terms of unit torque and the angular deflection.

In summary, the flexibility influence coefficients of a MDOF system can be determined as follows:

(1) Assume a unit load at point j ($j = 1$ to start with), due to this load, the displacements $\boldsymbol{x}_i =$ the jth column of $\boldsymbol{A}$ matrix.

(2) After completing step (1) for $j = 1$, the procedure is repeated for $j = 2, 3, \cdots, n$.

(3) The result is not as straightforward as in the previous case, after applying steps (1) and (2), the flexibility matrix $\boldsymbol{A}$ can be determined, then inversing the matrix $\boldsymbol{A}$ to find the stiffness matrix $\boldsymbol{K}$.

Example 5.2

Using the definitions given above, determine that the flexibility and stiffness matrices for a 2-DOF system shown in Fig. 5.17.

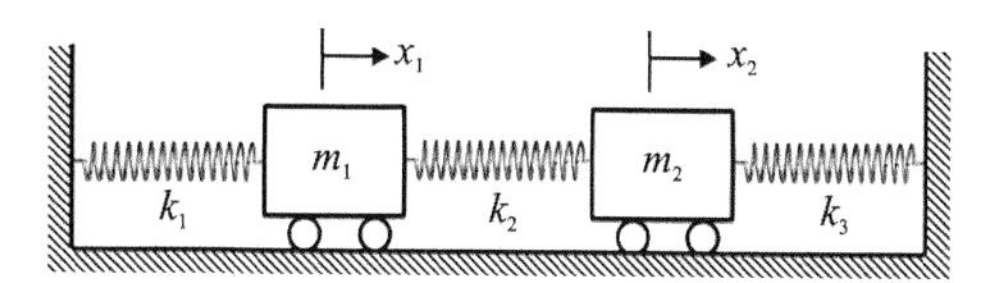

Fig. 5.17 A 2-DOF system

Solution:

Suppose that $x_1 = 1$ and $x_2 = 0$ (i. e. , give a unit displacement to m_1 while holding m_2 at its original position). Then k_{11} and k_{21} are the forces needed at location 1 and location 2, respectively, to maintain this static configuration. For this condition, it is clear that $f_1 = k_1 + k_2$ and $f_2 = -k_2$. Accordingly,

$$k_{11} = k_1 + k_2 \tag{5.51}$$

$$k_{21} = -k_2 \tag{5.52}$$

Similarly, suppose that $x_1 = 0$ and $x_2 = 1$. Then k_{12} and k_{22} are the forces needed at location 1 and location 2, respectively, to maintain the corresponding static configuration. It follows that

$$k_{21} = -k_2 \tag{5.53}$$

$$k_{22} = k_2 + k_3 \tag{5.54}$$

Consequently, the complete stiffness matrix can be expressed in terms of the stiffness elements in the system as follows:

$$\boldsymbol{K} = \begin{bmatrix} k_1 + k_2 & -k_2 \\ -k_2 & k_2 + k_3 \end{bmatrix} \tag{5.55}$$

Notice that flexibility matrix $\boldsymbol{A} = \boldsymbol{K}^{-1}$. So $\boldsymbol{A}$ matrix can be determined by using Eq. (5.55).

5.4 The Lagrange's Equation

The motion of particles and rigid bodies is governed by Newton's law. In this section, we will derive an alternate approach, placing Newton's law into a form particularly convenient for the MDOF systems. This approach results in a set of equations called Lagrange's equations. They are the beginning of a complex, more mathematical approach to mechanics called analytical dynamics. In this textbook we will only deal with this method at an elementary level. Lagrange's equations offer a systematic way to formulate the equations of motion of a mechanical MDOF system. A scalar approach is obtained by expressing the scalar quantities of kinetic and potential energy in terms of generalized coordinates.

Consideration of the energy in a dynamic system together with the use of the Lagrange's equation is a very powerful method of analysis for certain physically complex systems. It is an energy method which allows the equations of motion to be written in terms of any set of generalized coordinates. Generalized coordinates are a set of independent parameters which completely specify the system location and which are independent of any constraints. The fundamental form of Lagrange's equation can be written in terms of the generalized coordinates q_i as follows.

5.4.1 Derivation of Lagrange's equations in Cartesian coordinates

We begin by considering the conservation equations for a large number (n) of particles in a conservative force field using Cartesian coordinates of position x_i. For this system, the total kinetic energy can be expressed as

$$T=\sum_{i=1}^{n}\frac{1}{2}m_i\dot{x}_i^2 \tag{5.56}$$

where n is the number of degrees of freedom of the system.

The momentum of a given particle in a given direction can be obtained by differentiating this expression with respect to the appropriate x_i coordinate. This gives the momentum p_i for this particular particle in this coordinate direction.

$$p_i=\frac{\partial T}{\partial \dot{x}_i}=m_i\dot{x}_i \tag{5.57}$$

The time derivative of the momentum is

$$\frac{\mathrm{d}}{\mathrm{d}t}p_i=\frac{\mathrm{d}}{\mathrm{d}t}\left(\frac{\partial T}{\partial \dot{x}_i}\right)=m_i\ddot{x}_i \tag{5.58}$$

For a conservative force field, the force on a particle is given by the derivative of the potential energy U at the particle position in the desired direction.

$$F_i=-\frac{\partial U}{\partial x_i} \quad \text{and} \quad U=\sum_{i=1}^{n}\frac{1}{2}k_i x_i^2 \tag{5.59}$$

From Newton's law, and using Eqs. (5.57) and (5.58), we have

$$F_i=m_i\ddot{x}_i=\frac{\mathrm{d}p_i}{\mathrm{d}t}=\frac{\mathrm{d}}{\mathrm{d}t}\left(\frac{\partial T}{\partial \dot{x}_i}\right) \tag{5.60}$$

From Eqs. (5.59) and (5.60), we have

$$\frac{\mathrm{d}}{\mathrm{d}t}\left(\frac{\partial T}{\partial \dot{x}_i}\right) = -\frac{\partial U}{\partial x_i} \tag{5.61}$$

From Eqs. (5.56) and (5.59), it is clear that

$$\frac{\partial T}{\partial x_i} = 0 \quad \text{and} \quad \frac{\partial U}{\partial \dot{x}_i} = 0 \tag{5.62}$$

Using Eq. (5.62), we can rewrite Eq. (5.61) as follows:

$$\frac{\mathrm{d}}{\mathrm{d}t}\left[\frac{\partial (T-U)}{\partial \dot{x}_i}\right] - \frac{\partial (T-U)}{\partial x_i} = 0 \tag{5.63}$$

We now define $L = T - U$. L is called the **Lagrangian**. Eq. (5.63) takes the final form of Lagrange's equations in Cartesian coordinates.

$$\frac{\mathrm{d}}{\mathrm{d}t}\left(\frac{\partial L}{\partial \dot{x}_i}\right) - \frac{\partial L}{\partial x_i} = 0 \tag{5.64}$$

where i is taken over all of the degrees of freedom of the system.

Before moving on to more general coordinate systems, we will look at the application of Eq. (5.64) to a simple system.

Example 5.3

Recall the two-degree-of-freedom system shown in Fig. 5.17, find the motion equation by using Lagrange's equation.

Solution:

Notice that this is a two-degree-of-freedom system, governed by two differential equations. The governing equations could be obtained by applying Newton's law or the influence coefficient method.

The governing equations can also be obtained by direct application of Lagrange's equation. This approach is quite straight forward. Clearly, the kinetic energy T and the potential energy U can be expressed as follows:

$$T = \frac{1}{2}m_1\dot{x}_1^2 + \frac{1}{2}m_2\dot{x}_2^2$$

$$U = \frac{1}{2}k_1x_1^2 + \frac{1}{2}k_2(x_2 - x_1)^2 + \frac{1}{2}k_3x_2^2$$

Applying Lagrange's Eq. (5.64) to $L = T - U$, we get

$$\frac{\mathrm{d}}{\mathrm{d}t}\left(\frac{\partial L}{\partial \dot{x}_1}\right) - \frac{\partial L}{\partial x_1} = 0$$

$$\frac{\mathrm{d}}{\mathrm{d}t}\left(\frac{\partial L}{\partial \dot{x}_2}\right) - \frac{\partial L}{\partial x_2} = 0$$

We obtain the governing equations as follows:

$$m_1\ddot{x}_1 + k_1x_1 - k_2(x_2 - x_1) = 0$$

$$m_2\ddot{x}_2 + k_2(x_2 - x_1) + k_3x_2 = 0$$

or

$$\begin{bmatrix} m_1 & 0 \\ 0 & m_2 \end{bmatrix} \begin{bmatrix} \ddot{x}_1 \\ \ddot{x}_2 \end{bmatrix} + \begin{bmatrix} k_1+k_2 & -k_2 \\ -k_2 & k_2+k_3 \end{bmatrix} \begin{bmatrix} x_1 \\ x_2 \end{bmatrix} = 0$$

Clearly, for MDOF systems, this approach has advantages over the force balancing approach using Newton's law.

5.4.2 Extension to general coordinate systems

A significant advantage of the Lagrangian approach to developing equations of motion for complex systems comes as we leave the Cartesian coordinate system and move into a general coordinate system. We express the Cartesian variable x_i using generalized coordinates q_j.

$$x_i = x_i(q_1, \cdots, q_j, \cdots, q_n) \tag{5.65}$$

In the general case, each x_i could be dependent upon every q_j. What is remarkable about the Lagrange's formulation, is that Eq. (5.64) holds in a general coordinate system with x_i replaced by q_i.

$$\frac{\mathrm{d}}{\mathrm{d}t}\left(\frac{\partial L}{\partial \dot{q}_i}\right) - \frac{\partial L}{\partial q_i} = 0 \tag{5.66}$$

5.4.3 Application of Lagrange's equation

Considering a system with n degrees of freedom, generalized coordinates refer to any set of independent coordinates equal in number to the n degrees of freedom of the system under consideration. In this textbook, the generalized coordinates are denoted by q_i, $i = 1, 2, \cdots, n$, and are used to express the scalar notion of kinetic energy T and potential energy U.

Potential energy U of a vibration system typically only depends on the position of the system. Kinetic energy T typically depends on velocity, but may be also be position dependent. In terms of generalized coordinates q_i, $i = 1, 2, \cdots, n$, the scalar notion of kinetic energy T and potential energy U can be expressed as follows:

$$T = T(q_1, \cdots, q_n, \dot{q}_1, \cdots, \dot{q}_n) \tag{5.67a}$$

$$U = U(q_1, \cdots, q_n) \tag{5.67b}$$

Considering a conservative system, where all external and internal forces have a potential. In that case, the sum of kinetic energy T and potential energy U will be constant and the differential is equal to zero:

$$\mathrm{d}(T+U) = 0 \tag{5.68}$$

The above equation is basically a statement of the principle of conservation of energy. With the kinetic energy T and the potential energy U written as in Eq. (5.67), Lagrange's equation can be derived by summing up the kinetic and potential energy over all generalized coordinates q_i, $i = 1, 2, \cdots, n$.

Notice that Lagrangian $L = T-U$, and by using Eq. (5.67), Eq. (5.66) can be rewritten as follows:

$$\frac{\mathrm{d}}{\mathrm{d}t}\left[\frac{\partial(T-U)}{\partial \dot{q}_i}\right]-\frac{\partial(T-U)}{\partial q_i}=\frac{\mathrm{d}}{\mathrm{d}t}\left(\frac{\partial T}{\partial \dot{q}_i}\right)-\frac{\partial T}{\partial q_i}+\frac{\partial U}{\partial q_i}=0 \tag{5.69}$$

Eq. (5.69) constitutes Lagrange's equation for a conservative system. For systems that are nonconservative, Lagrange's equation in Eq. (5.69) can be generalized by including energy dissipation term and force term, such as

$$\frac{\mathrm{d}}{\mathrm{d}t}\left(\frac{\partial T}{\partial \dot{q}_i}\right)-\frac{\partial T}{\partial q_i}+\frac{\partial U}{\partial q_i}+\frac{\partial D}{\partial q_i}=\boldsymbol{Q}_i \tag{5.70}$$

where $\boldsymbol{Q}_i$ denotes the generalized forces acting on the system, and D is the energy dissipation function.

In summary, the equations of motion for a vibration system using Lagrange's equation can be determined as follows:

(1) Definition of the generalized coordinates q_i: This can be any set of independent coordinates equal in number to the n degrees of freedom of the system under consideration.

(2) Formulation of the kinetic energy T: The kinetic energy T can be written as

$$T=\frac{1}{2}\sum_{i=1}^{n}\sum_{j=1}^{n}M_{ij}\dot{q}_i\dot{q}_j=\frac{1}{2}\dot{\boldsymbol{q}}^{\mathrm{T}}\boldsymbol{M}\dot{\boldsymbol{q}} \tag{5.71}$$

(3) Formulation of the potential energy U: The potential energy U can be expressed as

$$U=\frac{1}{2}\sum_{i=1}^{n}\sum_{j=1}^{n}K_{ij}q_iq_j=\frac{1}{2}\boldsymbol{q}^{\mathrm{T}}\boldsymbol{K}\boldsymbol{q} \tag{5.72}$$

(4) Substituting Eqs. (5.71) and (5.72) into Lagrange's Eq. (5.69), and the motion equation can be obtained.

We also can determine the mass and stiffness matrices directly from Eqs. (5.71) and (5.72). From Eq. (5.71), the elements in mass matrix $\boldsymbol{M}$ can be obtained by

$$M_{ij}=\frac{\partial T^2}{\partial q_i\partial q_j} \tag{5.73}$$

Similarly, the elements in stiffness matrix $\boldsymbol{K}$ can be obtained by

$$K_{ij}=\frac{\partial^2 U}{\partial q_i\partial q_j} \tag{5.74}$$

From above analysis, we can find that the potential energy U is a quadratic function of the displacements, and the kinetic energy T is a quadratic function of the velocities. Notice that the kinetic energy T cannot be negative and vanishes only when all the velocities vanish. Eq. (5.71) is called as positive definite quadratic form and the mass matrix $\boldsymbol{M}$ is called as a positive definite matrix. On the other hand, the potential energy expression in Eq. (5.72) is a semi-positive definite quadratic form, the matrix $\boldsymbol{K}$ is positive definite only if the system is a stable one.

Example 5.4

Consider the three-mass system depicted in Fig. 5.18. Using Lagrange's equation, determine the equations of motion for 3-DOF system.

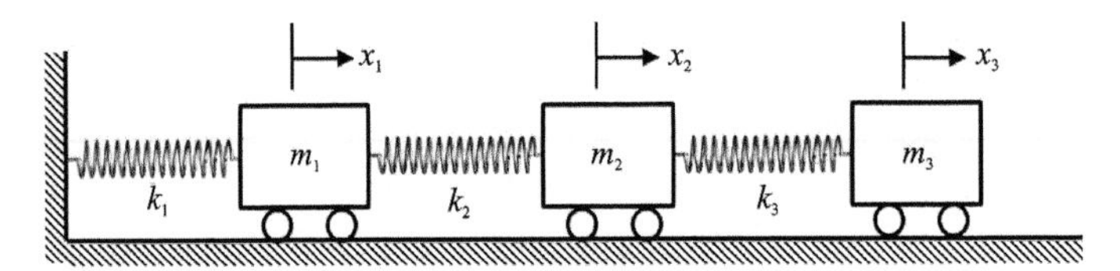

Fig. 5.18 Schematics of linear three-degree-of-freedom system

Solution:

As generalized coordinates for this 3-DOF system, simply the three (independent) positions x_i of the masses m_i, $i = 1, 2, 3$ can be chosen.

Clearly, the kinetic energy T and the potential energy U can be expressed as

$$T=\frac{1}{2}\sum_{k=1}^{3} m_k \dot{x}_k=\frac{1}{2}m_1\dot{x}_1+\frac{1}{2}m_2\dot{x}_2+\frac{1}{2}m_3\dot{x}_3$$

$$U=\frac{1}{2}k_1x_1^2+\frac{1}{2}k_2(x_2-x_1)^2+\frac{1}{2}k_3(x_3-x_2)^2$$

Apply the Lagrange's Eq. (5.69) with $x_i = q_i$, we get

$$m_1\dot{x}_1+(k_1+k_2)x_1-k_2x_2=0$$

$$m_2\dot{x}_2-k_2x_1+(k_2+k_3)x_2-k_3x_3=0$$

$$m_1\dot{x}_1+(k_1+k_2)x_1-k_2x_2=0$$

From the above equations, motion equation in matrix format can be written as

$$\begin{bmatrix} m_1 & 0 & 0 \\ 0 & m_2 & 0 \\ 0 & 0 & m_3 \end{bmatrix}\begin{bmatrix} \ddot{x}_1 \\ \ddot{x}_2 \\ \ddot{x}_3 \end{bmatrix}+\begin{bmatrix} k_1+k_2 & -k_2 & 0 \\ -k_2 & k_2+k_3 & -k_3 \\ 0 & -k_3 & k_3 \end{bmatrix}\begin{bmatrix} x_1 \\ x_2 \\ x_3 \end{bmatrix}=\begin{bmatrix} 0 \\ 0 \\ 0 \end{bmatrix} \tag{5.75}$$

An alternative way is based on Eqs. (5.73) and (5.74), it can be seen that M_{ij} for the mass matrix $\boldsymbol{M}$ and K_{ij} for the stiffness matrix $\boldsymbol{K}$ can be obtained

$$\boldsymbol{M}=\begin{bmatrix} m_1 & 0 & 0 \\ 0 & m_2 & 0 \\ 0 & 0 & m_3 \end{bmatrix},\quad \boldsymbol{K}=\begin{bmatrix} k_1+k_2 & -k_2 & 0 \\ -k_2 & k_2+k_3 & -k_3 \\ 0 & -k_3 & k_3 \end{bmatrix} \tag{5.76}$$

Clearly, the mass and stiffness matrices in Eq. (5.76) are the same as those in

Eq. (5.75).

Example 5.5

Consider the system depicted in Fig. 5.19, using Lagrange's equation, write the motion equation.

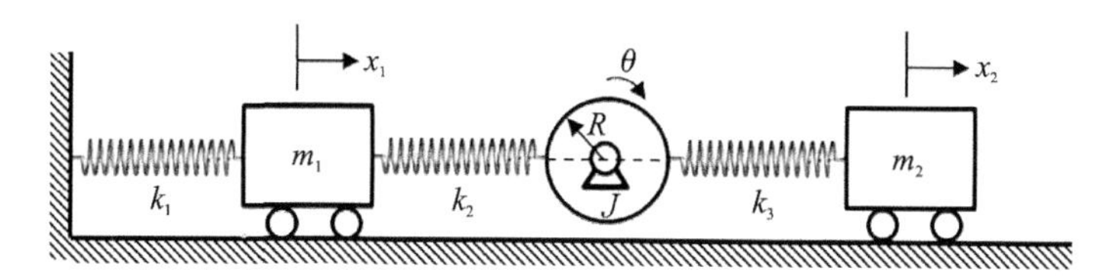

Fig. 5.19 Schematics of spring connected mass and inertia with external moment

Solution:

Clearly, the generalized coordinates are given as the linear displacement x_1 and x_2 for masses m_1 and m_2, respectively. And the inertia J can use the rotational displacement θ.

The kinetic energy is

$$T=\frac{1}{2}m\dot{x}_1^2+\frac{1}{2}m\dot{x}_2^2+\frac{1}{2}J\dot{\theta}^2$$

The potential energy is solely due to the presence of the three springs, we have

$$U=\frac{1}{2}k_1x_1^2+\frac{1}{2}k_2(R\theta-x_1)^2+\frac{1}{2}k_3(x_2-R\theta)^2$$

Combining the above information leads to the following three Lagrange's equations:

$$m_1\ddot{x}_1+(k_1+k_2)x_1-k_2R\theta=0$$

$$J\ddot{\theta}-k_2Rx_1+(k_2+k_3)R^2\theta-k_3Rx_2=0$$

$$m_2\ddot{x}_2-k_3R\theta+k_3x_2=0$$

Above equations can be combined in a matrix representation,

$$\begin{bmatrix} m_1 & 0 & 0 \\ 0 & J & 0 \\ 0 & 0 & m_2 \end{bmatrix}\begin{bmatrix} \ddot{x}_1 \\ \ddot{\theta} \\ \ddot{x}_2 \end{bmatrix}+\begin{bmatrix} k_1+k_2 & -k_2R & 0 \\ -k_2R & (k_2+k_3)R^2 & -k_3R \\ 0 & -k_3R & k_3 \end{bmatrix}\begin{bmatrix} x_1 \\ \theta \\ x_2 \end{bmatrix}=\begin{bmatrix} 0 \\ 0 \\ 0 \end{bmatrix}$$

Example 5.6

Consider the simplified model of a flexible four-story building depicted in Fig. 5.20. Assume that there is only horizontal displacement, using Lagrange's equation, write the motion equation.

Solution:

Clearly, the generalized coordinates can be chosen as the absolute horizontal displacement x_i.

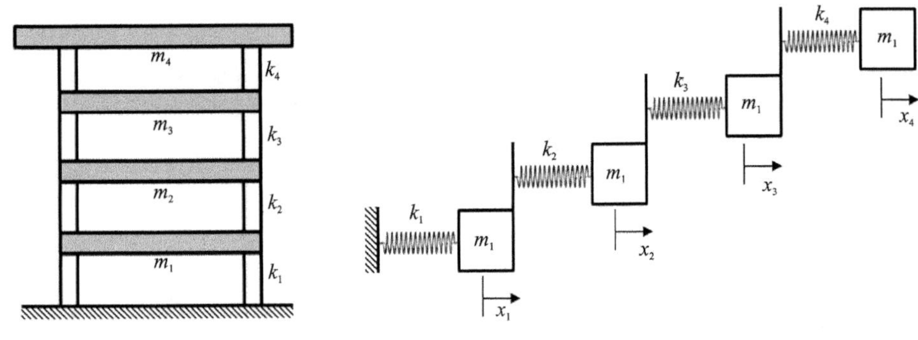

(a) Schematics of four-story building (b) Simple spring-mass model

Fig. 5.20 Schematics of four-story building and its simple spring-mass model

The kinetic energy T is determined by the total velocity and the mass of each floor,

$$T=\frac{1}{2}m_1\dot{x}_1^2+\frac{1}{2}m_2\dot{x}_2^2+\frac{1}{2}m_3\dot{x}_3^2++\frac{1}{2}m_4\dot{x}_4^2$$

Assuming linear shear-stiffness of the elastic side walls (as shown in Fig. 5.20 (b)), the potential energy U of the structure can be described in the generalized coordinates as follows:

$$U=\frac{1}{2}k_1x_1^2+\frac{1}{2}k_2(x_2-x_1)^2+\frac{1}{2}k_3(x_3-x_2)^2+\frac{1}{2}k_4(x_4-x_3)^2$$

The above information can be used for formulate the scalar Lagrange's equations, and which can be combined in a matrix representation:

$$\begin{bmatrix} m_1 & 0 & 0 & 0 \\ 0 & m_2 & 0 & 0 \\ 0 & 0 & m_3 & 0 \\ 0 & 0 & 0 & m_4 \end{bmatrix}\begin{bmatrix} \ddot{x}_1 \\ \ddot{x}_2 \\ \ddot{x}_3 \\ \ddot{x}_4 \end{bmatrix}+\begin{bmatrix} k_1+k_2 & -k_2 & 0 & 0 \\ -k_2 & k_2+k_3 & -k_3 & 0 \\ 0 & -k_3 & k_3+k_4 & -k_4 \\ 0 & 0 & -k_4 & k_4 \end{bmatrix}\begin{bmatrix} x_1 \\ x_2 \\ x_3 \\ x_4 \end{bmatrix}=\begin{bmatrix} 0 \\ 0 \\ 0 \\ 0 \end{bmatrix}$$

Above equation describes the dynamic behavior of the flexible four-story building.

5.5 Solving the Equations of Motion of Undamped Systems Using MATLAB

After we establish the motion equations for MDOF system, the next step is to solve these equations. Recall that the general form of the equation of motion for a vibrating system is

$$\boldsymbol{M}\ddot{\boldsymbol{x}}+\boldsymbol{C}\dot{\boldsymbol{x}}+\boldsymbol{K}\boldsymbol{x}=\boldsymbol{f} \tag{5.77}$$

where $\boldsymbol{x}$ is a time dependent vector that describes the motion, $\boldsymbol{M}$, $\boldsymbol{C}$ and $\boldsymbol{K}$ are mass, damping and stiffness matrices.

For undamped free vibration case, we set $\boldsymbol{C} = \boldsymbol{0}$ and $\boldsymbol{f} = \boldsymbol{0}$. So Eq. (5.77) can be simplified as follows:

$$\boldsymbol{M}\ddot{\boldsymbol{x}} + \boldsymbol{K}\boldsymbol{x} = \boldsymbol{0} \tag{5.78}$$

Since we are interested in finding harmonic solutions for $\boldsymbol{x}$, we can simply assume that the solution has the form as follows:

$$\boldsymbol{x} = \boldsymbol{X}\sin(\omega t) \tag{5.79}$$

Substituting Eq. (5.79) into Eq. (5.78), we get

$$-\boldsymbol{M}\boldsymbol{X}\omega^2\sin(\omega t) + \boldsymbol{K}\boldsymbol{X}\sin(\omega t) = 0 \tag{5.80}$$

From Eq. (5.80), since the solution must be valid for all time, the sine term will generally be nonzero, so it is easy to find that

$$\lambda\boldsymbol{M}\boldsymbol{X} = \boldsymbol{K}\boldsymbol{X} \quad \text{with} \quad \lambda = \omega^2 \tag{5.81a}$$

or

$$(\boldsymbol{K} - \lambda\boldsymbol{M})\boldsymbol{X} = \boldsymbol{0} \tag{5.81b}$$

From Eq. (5.81), we can find that the determinant of the coefficients matrix must be zero for a nontrivial solution of the vectors $\boldsymbol{X}$. So we get

$$\det(\boldsymbol{K} - \lambda\boldsymbol{M}) = 0 \tag{5.82}$$

In Eq. (5.81), the vectors $\boldsymbol{X}$ and λ scalars that satisfy a matrix equation of the form $\lambda\boldsymbol{M}\boldsymbol{X} = \boldsymbol{K}\boldsymbol{X}$ are called "generalized eigenvectors" and "generalized eigenvalues" of the equation. And Eq. (5.82) is the frequency or characteristic equation. If Eq. (5.82) is solved, the natural frequencies can be obtained. However, it is very difficult to find exact formulas for $\boldsymbol{X}$ and λ for a large matrix by hand. But MATLAB has built-in functions that will compute generalized eigenvectors and eigenvalues given numerical values for $\boldsymbol{M}$ and $\boldsymbol{K}$. The special values of λ satisfying $\lambda\boldsymbol{M}\boldsymbol{X} = \boldsymbol{K}\boldsymbol{X}$ are related to the natural frequencies by $\omega_i = \sqrt{\lambda_i}$. The special vectors $\boldsymbol{X}_k$ are the kth mode shape of the system. These are the special initial displacements that will cause the mass to vibrate harmonically. It is easy to verify that two eigenvectors (mode shapes) corresponding to two different eigenvalues are mutually orthogonal with respect to the mass and stiffness matrices.

If we only want to know the natural frequencies, we can use the MATLAB command as follows:

```
d = eig(K,M)
```

This returns a vector $\boldsymbol{d}$, containing all the values of λ satisfying $\lambda\boldsymbol{M}\boldsymbol{X} = \boldsymbol{K}\boldsymbol{X}$ (for an $n \times n$ matrix, there are usually n different values). The natural frequencies follow as $\omega_i = \sqrt{\lambda_i}$.

If we want to find both the eigenvalues and eigenvectors, we must use

```
[V,D] = eig(K,M)
```

This returns two matrices, $\boldsymbol{V}$ and $\boldsymbol{D}$. Each column of the matrix $\boldsymbol{V}$ corresponds to a vector $\boldsymbol{X}$ that satisfies the equation, and the diagonal elements of $\boldsymbol{D}$ contain the corresponding value of λ. To extract the ith natural frequency and mode shape, use the following MATLAB command:

```
omega = sqrt(D(i,i))
X = V(:,i)
```

Example 5.7

Recall the free vibration problem shown in Fig. 5.1(a), it has been shown that the mass and stiffness matrices are (see Eq. (5.3))

$$\boldsymbol{M}=\begin{bmatrix} m_1 & 0 \\ 0 & m_2 \end{bmatrix} \quad \text{and} \quad \boldsymbol{K}=\begin{bmatrix} k_1+k_2 & -k_2 \\ -k_2 & k_2 \end{bmatrix}$$

Assume that $m_1 = m_2 = 1$ kg, $k_1 = k_2 = 1$ N/s, calculate the natural frequencies and mode shapes.

Solution:

The following MATLAB codes can be used to automatically compute the natural frequencies and mode shapes.

```
m1=1;
m2=1;
k1=1;
k2=1;
M=[m1 0; 0 m2];
K=[k1+k2 -k2; -k2 k2];
[V,D]=eig(K,M);
for i=1 :2
display(['The ' num2str(i) ' mode'])
omega = sqrt(D(i,i))
X = V(:,i)
end
```

The following results are obtained:

```
The 1 mode
omega =
    0.6180
X =
```

```
    -0.5257
    -0.8507
The 2 mode
omega =
    1.6180
X =
    -0.8507
0.5257
```

Clearly, this gives the natural frequencies as $\omega_1 = 0.618$ rad/s and $\omega_1 = 1.618$ rad/s. The corresponding mode shapes as $X_1 = (-0.5257, -0.8507)$ (both masses displace in the same direction, and their ratio is 1.618) and $X_2 = (0.8507, -0.5257)$ (the two masses displace in opposite directions, and their ratio is -0.618). And we get the same results as solving the characteristics equation (see Eqs. (5.10) and (5.11).

Proportionality of the mode shape

A mode shape is represented by a vector, called eigenvector mathematically. The absolute magnitude of individual elements of an eigenvector is indefinite —an eigenvector multiplied by a nonzero scalar is still an eigenvector. But the **relative proportion** of the individual elements of an eigenvector is fixed. For this reason, an eigenvector is normally normalized (scaled according to a rule) and then becomes unique. Normalization helps to graphically display a mode, compare modes and facilitate computation.

There are several ways to normalize the mode shapes:

(1) Let the maximum element be one and the other elements scaled as $X'_j(k) = \dfrac{X_j(k)}{\max(X_j)}$ for $j, k = 1, 2, \cdots, n$.

(2) Multiply a factor to an eigenvector so that $\boldsymbol{X}_j^{\mathrm{T}}\boldsymbol{M}\boldsymbol{X}_k = \delta_{jk} = \begin{cases} 1 & j = k \\ 0 & j \neq k \end{cases}$, it is called as mass normalization.

(3) Scale an eigenvector so that $\sqrt{\sum_{k=1}^{n} X_j^2(k)} = 1$.

A normalized eigenvector is called a normal eigenvector (or normal mode). And in MATLAB, the mass normalization will be automatically obtained by using function eig.

5.6 Vibration of Undamped MDOF System

Throughout the rest of this section, we will focus on exploring the behavior of systems of springs and masses. This is not because spring/mass systems are of any particular interest, but because they are easy to visualize, and, more importantly the equations of motion for a spring-mass system are identical to those of any linear system.

5.6.1 Free response of undamped MDOF systems

After we obtain the natural frequencies and mode shapes of an n degree-of-freedom undamped system, the next step is to obtain the free response of the system. According to Eq. (5.78), the differential equation of undamped system is

$$\boldsymbol{M}\ddot{\boldsymbol{x}}+\boldsymbol{K}\boldsymbol{x}=\boldsymbol{0} \tag{5.83a}$$

and the initial conditions are

$$\left.\begin{aligned}\boldsymbol{x}(0)&=\boldsymbol{x}_0\\ \dot{\boldsymbol{x}}(0)&=\boldsymbol{v}_0\end{aligned}\right\} \tag{5.83b}$$

The response $\boldsymbol{x}(t)$ in this case can be obtained without decoupling the equations of motion (5.83a), as done in the previous section.

Notice that Eq. (5.83) can be solved by assuming $\boldsymbol{x}=\boldsymbol{X}\sin(\omega t)$, we get

$$(\boldsymbol{K}-\lambda_i\boldsymbol{M})\boldsymbol{X}_i=\boldsymbol{0}\quad \text{with}\quad \lambda_i=\omega_i^2 \tag{5.84}$$

Case 1: $\lambda_i\neq 0$

Clearly, if $\lambda_i\neq 0$, then the vector

$$a_i\boldsymbol{X}_i\sin(\sqrt{\lambda_i}t)\quad \text{and}\quad b_i\boldsymbol{X}_i\cos(\sqrt{\lambda_i}t) \tag{5.85}$$

satisfy the differential Eq. (5.83), and a_i and b_i are the arbitrary nonzero constant.

Proof:

This statement is easy to be proofed. Substituting Eq. (5.85) for $\boldsymbol{x}$ in Eq. (5.83a) and using Eq. (5.84), we can obtain

$$\begin{aligned}\boldsymbol{M}\ddot{\boldsymbol{x}}+\boldsymbol{K}\boldsymbol{x}&=-a_i\lambda_i\boldsymbol{M}\boldsymbol{X}_i\sin(\sqrt{\lambda_i}t)+a_i\boldsymbol{K}\boldsymbol{X}_i\sin(\sqrt{\lambda_i}t)\\&=a_i\sin(\sqrt{\lambda_i}t)(\boldsymbol{K}-\lambda_i\boldsymbol{M})\boldsymbol{X}_i=\boldsymbol{0}\\ \boldsymbol{M}\ddot{\boldsymbol{x}}+\boldsymbol{K}\boldsymbol{x}&=-b_i\lambda_i\boldsymbol{M}\boldsymbol{X}_i\cos(\sqrt{\lambda_i}t)+b_i\boldsymbol{K}\boldsymbol{X}_i\cos(\sqrt{\lambda_i}t)\\&=b_i\cos(\sqrt{\lambda_i}t)(\boldsymbol{K}-\lambda_i\boldsymbol{M})\boldsymbol{X}_i=\boldsymbol{0}\end{aligned}$$

It is easy to find that the sum of these solutions in Eq. (5.85) is also a solution to Eq. (5.83), yields

$$x(t)=\sum_{i=1}^{n}\left[a_i\boldsymbol{X}_i\sin(\sqrt{\lambda_i}t)+b_i\boldsymbol{X}_i\cos(\sqrt{\lambda_i}t)\right] \tag{5.86}$$

Clearly, Eq. (5.86) satisfies the differential Eq. (5.83a). So $x(t)$ in Eq. (5.86) can be expressed in terms of $2n$ arbitrary constants and it is the general solution of Eq. (5.83).

The coefficients a_i and b_i are determined by the initial conditions in Eq. (5.83b). Using Eq. (5.86), the initial displacement vector can be written as

$$\boldsymbol{x}(0)=\sum_{i=1}^{n} b_i \boldsymbol{X}_i=\boldsymbol{X}\boldsymbol{b}=\boldsymbol{x}_0 \tag{5.87}$$

where the modal matrix $\boldsymbol{X}$ is given in the previous section,

$$\boldsymbol{X}=[\boldsymbol{X}_1, \boldsymbol{X}_2, \cdots, \boldsymbol{X}_n], \text{ and } \boldsymbol{b}=[b_1, b_2, \cdots, b_n]^{\mathrm{T}}$$

From Eq. (5.87), We can obtain

$$\boldsymbol{b}=\boldsymbol{X}^{-1}\boldsymbol{x}_0 \tag{5.88}$$

Then substituting the initial velocity condition Eq. (5.83b) into Eq. (5.86), gives

$$\dot{\boldsymbol{x}}(0)=\sum_{i=1}^{n} a_i \sqrt{\lambda_i}\boldsymbol{X}_i=\boldsymbol{X}\boldsymbol{\Lambda}\boldsymbol{a}=\boldsymbol{v}_0 \tag{5.89}$$

where $\boldsymbol{\Lambda}$ is a diagonal matrix, with diagonal element $\sqrt{\lambda_i}$, and $\boldsymbol{a}=[a_1, a_2, \cdots, a_n]^{\mathrm{T}}$.

We thus have

$$\boldsymbol{a}=\boldsymbol{\Lambda}^{-1}\boldsymbol{X}^{-1}\boldsymbol{v}_0 \tag{5.90}$$

where $\boldsymbol{\Lambda}^{-1}$ is a diagonal matrix, with diagonal element $\dfrac{1}{\sqrt{\lambda_i}}$.

In summary, the general solution of Eq. (5.83) is given by Eq. (5.86), where the coefficients a_i and $b_i (i=1, 2, \cdots, n)$ are determined by Eqs. (5.88) and (5.90), respectively. The following example demonstrates this result.

Example 5.8

Assume there is a two-degree-of-freedom system with $m_1=m_2=1$ kg, $k_1=k_3=1$ N/s and $k_2=2$ N/s, as shown in Fig. 5.21. Determine the response of the system if the initial conditions are

$$\boldsymbol{x}(0)=\begin{bmatrix}1\\-1\end{bmatrix}, \quad \dot{\boldsymbol{x}}(0)=\boldsymbol{v}(0)=\begin{bmatrix}1\\0\end{bmatrix}$$

Fig. 5.21 Free vibration of a two-degree-of-freedom system

Solution:

The differential equation of motion is

$$\boldsymbol{M}\ddot{\boldsymbol{x}} + \boldsymbol{K}\boldsymbol{x} = \boldsymbol{0}$$

It is easy to find that the mass and stiffness matrices are

$$\boldsymbol{M} = \begin{bmatrix} m_1 & 0 \\ 0 & m_2 \end{bmatrix} = \begin{bmatrix} 1 & 0 \\ 0 & 1 \end{bmatrix}, \quad \boldsymbol{K} = \begin{bmatrix} k_1 + k_2 & -k_2 \\ -k_2 & k_2 + k_3 \end{bmatrix} = \begin{bmatrix} 3 & -2 \\ -2 & 3 \end{bmatrix}$$

According to Section 5.6, using MATLAB function eig, it was found that the spectral and modal matrices for this system are

$$\boldsymbol{\Lambda} = \begin{bmatrix} 1 & 0 \\ 0 & 5 \end{bmatrix}, \quad \boldsymbol{X} = \begin{bmatrix} -0.707 & -0.707 \\ -0.707 & 0.707 \end{bmatrix} = [\boldsymbol{X}_1 \quad \boldsymbol{X}_2]$$

By using Eqs. (5.88) and (5.90), we get

$$\boldsymbol{a} = \boldsymbol{\Lambda}^{-1}\boldsymbol{X}^{-1}\boldsymbol{v}_0 = \begin{bmatrix} 1 & 0 \\ 0 & \dfrac{1}{\sqrt{5}} \end{bmatrix} \begin{bmatrix} -0.707 & -0.707 \\ -0.707 & 0.707 \end{bmatrix} \begin{bmatrix} 1 \\ 0 \end{bmatrix} = \begin{bmatrix} -0.707 \\ -0.3162 \end{bmatrix}$$

$$\boldsymbol{b} = \boldsymbol{X}^{-1}\boldsymbol{x}_0 = \begin{bmatrix} -0.707 & -0.707 \\ -0.707 & 0.707 \end{bmatrix} \begin{bmatrix} 1 \\ -1 \end{bmatrix} = \begin{bmatrix} 0 \\ -1.414 \end{bmatrix}$$

By using Eq. (5.86), the solution can be obtained as

$$\begin{aligned} x(t) &= \sum_{i=1}^{n} \left[a_i \boldsymbol{X}_i \sin(\sqrt{\lambda_i} t) + b_i \boldsymbol{X}_i \cos(\sqrt{\lambda_i} t)\right] \\ &= -0.707 \begin{bmatrix} -0.707 \\ -0.707 \end{bmatrix} \sin t - 0.3162 \begin{bmatrix} -0.707 \\ 0.707 \end{bmatrix} \sin\sqrt{5}t - 1.414 \begin{bmatrix} -0.707 \\ 0.707 \end{bmatrix} \cos\sqrt{5}t \\ &= \begin{bmatrix} 0.5 \\ 0.5 \end{bmatrix} \sin t + \begin{bmatrix} 0.224 \\ -0.224 \end{bmatrix} \sin\sqrt{5}t + \begin{bmatrix} 1 \\ -1 \end{bmatrix} \cos\sqrt{5}t \end{aligned}$$

Thus we get

$$\begin{cases} x_1(t) = 0.5\sin t + 0.224\sin\sqrt{5}t + \cos\sqrt{5}t \\ x_2(t) = 0.5\sin t - 0.224\sin\sqrt{5}t - \cos\sqrt{5}t \end{cases}$$

The responses can be plotted as in Fig. 5.22. It should be noticed that the responses of the free vibration system depend on the initial conditions, if the initial displacement condition is proportional to the i th mode shape, the responses will be vibrated with the corresponding natural frequencies. For example, if the initial conditions are changed as $\boldsymbol{x}(0) = [1,1]^{\mathrm{T}}$, $\boldsymbol{v}(0) = [0,0]^{\mathrm{T}}$, two masses vibrate in phase ($x_1(t)/x_2(t) = 1$) at 1 rad/s (the first natural frequency). If the initial conditions are changed as $\boldsymbol{x}(0) = [1,-1]^{\mathrm{T}}$, $\boldsymbol{v}(0) = [0,0]^{\mathrm{T}}$. two masses vibrate out of phase ($x_1(t)/x_2(t) = -1$) at 2.236 rad/s (the second natural frequency), as shown in Fig. 5.23.

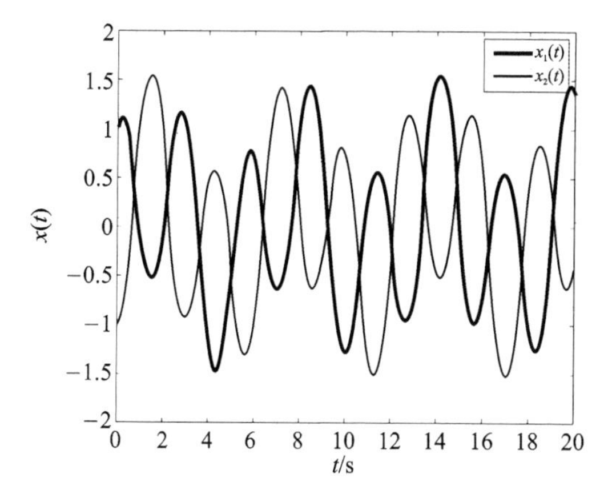

Fig. 5.22 The time response for Example 5.8

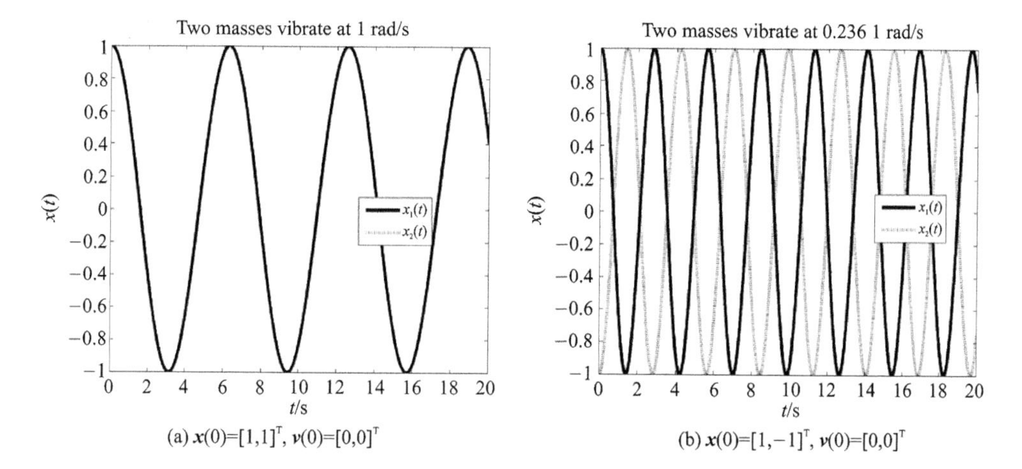

Fig. 5.23 The time response for Example 5.8 when the initial conditions are changed

Case 2: $\lambda_1 = 0$

We note that if $\lambda_1 = 0$, then $a_1 \boldsymbol{X}_1 \sin(\sqrt{\lambda_1}\, t) \equiv 0$. It means that there are only $2n-1$ arbitrary constants in Eq. (5.86). We thus conclude that in this case Eq. (5.86) is not the general solution of Eq. (5.83). However

$$(a_1 t + b_1)\boldsymbol{X}_1 \tag{5.91}$$

for arbitrary scalar a_1 and b_1 does solve Eq. (5.83) in this case.

Proof:

If $\lambda_1 = 0$, then Eq. (5.84) implies that

$$\boldsymbol{K}\boldsymbol{X}_1 = 0 \tag{5.92}$$

Substituting Eq. (5.91) into Eq. (5.83) and using Eq. (5.92), we have

$$\boldsymbol{M}\ddot{\boldsymbol{x}} + \boldsymbol{K}\boldsymbol{x} = (a_1 t + b_1)\boldsymbol{K}\boldsymbol{X}_1 = 0$$

This completes the proof of the above statement.

Therefore, when $\lambda_1 = 0$, the general solution of Eq. (5.83) can be expressed as

$$x(t) = a_1 t\boldsymbol{X}_1 + b_1\boldsymbol{X}_1 + \sum_{i=2}^{n}\left[a_i\boldsymbol{X}_i\sin(\sqrt{\lambda_i}t) + b_i\boldsymbol{X}_i\cos(\sqrt{\lambda_i}t)\right] \tag{5.93}$$

We see that there is one non-oscillatory mode of motion, namely $a_1 t\boldsymbol{X}_1$, in which the system moves from its static equilibrium position without bound as t increases. This mode of motion is called a rigid body mode of motion. Clearly the system can have such mode of motion only if there is no spring connection between the system and the ground. In this case $\lambda_1 = 0$ and the response of the system is determined by Eq. (5.93). The $2n$ coefficients in Eq. (5.93) are determined by the initial condition. The $b_i(i=1, 2,\cdots, n)$ can be obtained by using Eq. (5.88). However, the a_i cannot be obtained by Eq. (5.90), because $\boldsymbol{\Lambda}$ is singular and $\boldsymbol{\Lambda}^{-1/2}$ does not exist for $\lambda_1 = 0$. We may obtain these coefficients directly from the initial conditions, it will be demonstrated by the following example.

Example 5.9

Assume there is a three-degree-of-freedom system with $m_1 = m_3 = 1$ kg, $m_2 = 2$ kg, $k_1 = k_2 = 5$ N/s, as shown in Fig. 5.24. Determine the response of the system if the initial conditions are

$$\boldsymbol{x}(0) = \begin{bmatrix} x_1(0) \\ x_2(0) \\ x_3(0) \end{bmatrix} = \begin{bmatrix} x_0 \\ 0 \\ 0 \end{bmatrix}, \quad \dot{\boldsymbol{x}}(0) = \begin{bmatrix} v_1(0) \\ v_2(0) \\ v_3(0) \end{bmatrix} = \begin{bmatrix} v_0 \\ 0 \\ 0 \end{bmatrix}$$

Fig. 5.24 Free vibration of a three-degree-of-freedom system

Solution:

The matrix differential equation is

$$\boldsymbol{M}\ddot{\boldsymbol{x}} + \boldsymbol{K}\boldsymbol{x} = \boldsymbol{0}$$

where

$$\boldsymbol{M} = \begin{bmatrix} 1 & 0 & 0 \\ 0 & 2 & 0 \\ 0 & 0 & 1 \end{bmatrix}, \quad \boldsymbol{K} = \begin{bmatrix} 5 & -5 & 0 \\ -5 & 10 & -5 \\ 0 & -5 & 5 \end{bmatrix}$$

By using MATLAB function eig, as shown in Section 5.6, the $\boldsymbol{\Lambda}$ and $\boldsymbol{X}$ matrices for this system are

$$\boldsymbol{\Lambda}=\begin{bmatrix}0 & 0 & 0\\0 & 5 & 0\\0 & 0 & 10\end{bmatrix},\quad \boldsymbol{X}=\begin{bmatrix}0.5 & -0.707 & -0.5\\0.5 & 0 & 0.5\\0.5 & 0.707 & -0.5\end{bmatrix}=[\boldsymbol{X}_1\quad \boldsymbol{X}_2\quad \boldsymbol{X}_3]$$

Notice that $\lambda_1=0$, so the general solution for this problem is given by Eq. (5.93),

$$x(t)=a_1t\boldsymbol{X}_1+b_1\boldsymbol{X}_1+\sum_{i=2}^{3}\left[a_i\boldsymbol{X}_i\sin(\sqrt{\lambda_i}t)+b_i\boldsymbol{X}_i\cos(\sqrt{\lambda_i}t)\right]\tag{5.94}$$

By using Eq. (5.88), we get

$$\boldsymbol{b}=\boldsymbol{X}^{-1}\boldsymbol{x}_0=\begin{bmatrix}0.5 & -0.707 & -0.5\\0.5 & 0 & 0.5\\0.5 & 0.707 & -0.5\end{bmatrix}\begin{bmatrix}x_0\\0\\0\end{bmatrix}=\begin{bmatrix}0.5\\-0.707\\-0.5\end{bmatrix}x_0$$

Thus the general solution in Eq. (5.94) can be rewritten as

$$\boldsymbol{x}(t)=a_1t\begin{bmatrix}0.5\\0.5\\0.5\end{bmatrix}+\begin{bmatrix}0.25\\0.25\\0.25\end{bmatrix}x_0+a_2\begin{bmatrix}-0.707\\0\\0.707\end{bmatrix}\sin\sqrt{5}t-$$

$$\begin{bmatrix}-0.5\\0\\0.5\end{bmatrix}x_0\cos\sqrt{5}t+a_3\begin{bmatrix}-0.5\\0.5\\-0.5\end{bmatrix}\sin\sqrt{10}t-\begin{bmatrix}-0.25\\0.25\\-0.25\end{bmatrix}x_0\cos\sqrt{10}t\tag{5.95}$$

Notice that there are three unknown scalars (a_1, a_2 and a_3) in Eq. (5.95), differentiating Eq. (5.95), with respect to the time t, the velocity can be obtained as

$$\dot{\boldsymbol{x}}(t)=a_1\begin{bmatrix}0.5\\0.5\\0.5\end{bmatrix}+a_2\sqrt{5}\begin{bmatrix}-0.707\\0\\0.707\end{bmatrix}\cos\sqrt{5}t-\sqrt{5}x_0\begin{bmatrix}-0.5\\0\\0.5\end{bmatrix}\sin\sqrt{5}t+$$

$$a_3\sqrt{10}\begin{bmatrix}-0.5\\0.5\\-0.5\end{bmatrix}\cos\sqrt{10}t+\sqrt{10}\begin{bmatrix}-0.25\\0.25\\-0.25\end{bmatrix}x_0\sin\sqrt{10}t\tag{5.96}$$

At $t=0$, Eq. (5.96) is equal to the initial velocity condition, that is

$$\dot{\boldsymbol{x}}(0)=\begin{bmatrix}v_0\\0\\0\end{bmatrix}=a_1\begin{bmatrix}0.5\\0.5\\0.5\end{bmatrix}+a_2\begin{bmatrix}-0.707\\0\\0.707\end{bmatrix}\sqrt{5}+a_3\sqrt{10}\begin{bmatrix}-0.5\\0.5\\-0.5\end{bmatrix}\tag{5.97}$$

Eq. (5.97) can be further rewritten into matrix form:

$$\begin{bmatrix}v_0\\0\\0\end{bmatrix}=\begin{bmatrix}0.5 & -0.707\sqrt{5} & -0.5\sqrt{10}\\0.5 & 0 & 0.5\sqrt{10}\\0.5 & 0.707\sqrt{5} & -0.5\sqrt{10}\end{bmatrix}\begin{bmatrix}a_1\\a_2\\a_3\end{bmatrix}$$

Thus

$$\begin{bmatrix}a_1\\a_2\\a_3\end{bmatrix}=\mathrm{inv}\left(\begin{bmatrix}0.5 & -0.707\sqrt{5} & -0.5\sqrt{10}\\0.5 & 0 & 0.5\sqrt{10}\\0.5 & 0.707\sqrt{5} & -0.5\sqrt{10}\end{bmatrix}\right)\begin{bmatrix}v_0\\0\\0\end{bmatrix}=\begin{bmatrix}0.5\\-0.316\\-0.158\end{bmatrix}v_0$$

where inv( ) denotes the inverse of the matrix.

Substituting a_i into Eq. (5.95), we can obtain the system response:

$$\boldsymbol{x}(t)=\begin{bmatrix}0.25\\0.25\\0.25\end{bmatrix}v_0t+\begin{bmatrix}0.25\\0.25\\0.25\end{bmatrix}x_0+\begin{bmatrix}0.224\\0\\-0.224\end{bmatrix}v_0\sin\sqrt{5}t-$$

$$\begin{bmatrix}-0.5\\0\\0.5\end{bmatrix}x_0\cos\sqrt{5}t-\begin{bmatrix}-0.079\\0.079\\-0.079\end{bmatrix}v_0\sin\sqrt{10}t-\begin{bmatrix}-0.25\\0.25\\-0.25\end{bmatrix}x_0\cos\sqrt{10}t \tag{5.98}$$

Rewrite Eq. (5.98) in component form, we get

$$x_1(t)=0.25v_0t+0.25x_0+0.224v_0\sin\sqrt{5}t+0.5x_0\cos\sqrt{5}t+$$
$$0.079v_0\sin\sqrt{10}t+0.25x_0\cos\sqrt{10}t \tag{5.99a}$$

$$x_2(t)=0.25v_0t+0.25x_0-0.079v_0\sin\sqrt{10}t-0.25x_0\cos\sqrt{10}t \tag{5.99b}$$

$$x_3(t)=0.25v_0t+0.25x_0-0.224v_0\sin\sqrt{5}t-0.5x_0\cos\sqrt{5}t+$$
$$0.079v_0\sin\sqrt{10}t+0.25x_0\cos\sqrt{10}t \tag{5.99c}$$

From Eq. (5.99), it can be found that the system moves to right according to its rigid bode mode of motion $0.25v_0t$. It means that the system has a constant velocity $0.25v_0$. There are two additional components of vibrations, with frequency $\sqrt{5}$ rad/s and $\sqrt{10}$ rad/s, respectively. The rigid body mode of motion is because the system has NO spring connections with ground. Such system is called as a free-free system.

We note that in the analysis of this section no assumption regarding the symmetry of $\boldsymbol{M}$ and $\boldsymbol{K}$ has been made. We can conclude that Eqs. (5.86) and (5.93) represent general solutions regardless of measurements origin. The following example demonstrates the use of the general solution in Eq. (5.86) for non-symmetric mass and stiffness matrices.

Example 5.10

Recall the double pendulum in Fig. 5.12, according to Eqs. (5.34) and (5.40a), the differential equations of motion for the double pendulum can be written in different matrix forms as follows:

$$\begin{bmatrix}m_1L_1 & 0\\ L_1 & L_2\end{bmatrix}\begin{bmatrix}\ddot{\theta}_1\\ \ddot{\theta}_2\end{bmatrix}+\begin{bmatrix}(m_1+m_2)g & -m_2g\\ 0 & g\end{bmatrix}\begin{bmatrix}\theta_1\\ \theta_2\end{bmatrix}=\begin{bmatrix}0\\0\end{bmatrix}$$

Assume the system if the initial conditions of the double pendulum are

$$\begin{bmatrix}\theta_1(0)\\ \theta_2(0)\end{bmatrix}=\begin{bmatrix}0\\0.01\end{bmatrix},\quad \begin{bmatrix}\dot{\theta}_1(0)\\ \dot{\theta}_2(0)\end{bmatrix}=\begin{bmatrix}0\\0\end{bmatrix}$$

and assume $m_1=1$ kg, $m_2=2$ kg, $L_1=L_2=1$ m. Determine the response of the system.

Solution:

The differential equations of motion for the double pendulum for our case can be written as

$$\begin{bmatrix} 1 & 0 \\ 1 & 1 \end{bmatrix} \begin{bmatrix} \ddot{\theta}_1 \\ \ddot{\theta}_2 \end{bmatrix} + \begin{bmatrix} 3g & -2g \\ 0 & g \end{bmatrix} \begin{bmatrix} \theta_1 \\ \theta_2 \end{bmatrix} = \begin{bmatrix} 0 \\ 0 \end{bmatrix}$$

Similar to previous Examples, by using MATLAB function eig, the $\boldsymbol{\Lambda}$ and $\boldsymbol{X}$ matrices for this system can be solved, that is

$$\boldsymbol{\Lambda} = \begin{bmatrix} 0.551 & 0 \\ 0 & 5.45 \end{bmatrix} g, \quad \boldsymbol{X} = \begin{bmatrix} -0.817 & -0.817 \\ -1 & 1 \end{bmatrix}$$

By using Eqs. (5.88) and (5.90), we get

$$\boldsymbol{a} = \boldsymbol{\Lambda}^{-1}\boldsymbol{X}^{-1}\boldsymbol{v}_0 = \begin{bmatrix} 0 \\ 0 \end{bmatrix}, \quad \boldsymbol{b} = \boldsymbol{X}^{-1}\boldsymbol{x}_0 = \begin{bmatrix} -0.005 \\ 0.005 \end{bmatrix}$$

By using Eq. (5.86), the solution can be obtained as

$$\begin{aligned} \boldsymbol{\theta}(t) &= \sum_{i=1}^{n} \left[a_i \boldsymbol{X}_i \sin(\sqrt{\lambda_i} t) + b_i \boldsymbol{X}_i \cos(\sqrt{\lambda_i} t) \right] \\ &= -0.005 \begin{bmatrix} -0.817 \\ -1 \end{bmatrix} \cos(0.551gt) + 0.005 \begin{bmatrix} -0.817 \\ 1 \end{bmatrix} \cos(5.45gt) \end{aligned}$$

Rewrite above equation, we get

$$\begin{cases} \theta_1(t) = 0.00408\cos(0.551gt) - 0.00408\cos(5.45gt) \\ \theta_2(t) = 0.005\cos(0.551gt) + 0.005\cos(5.45gt) \end{cases}$$

These responses can be plotted as in Fig. 5.25.

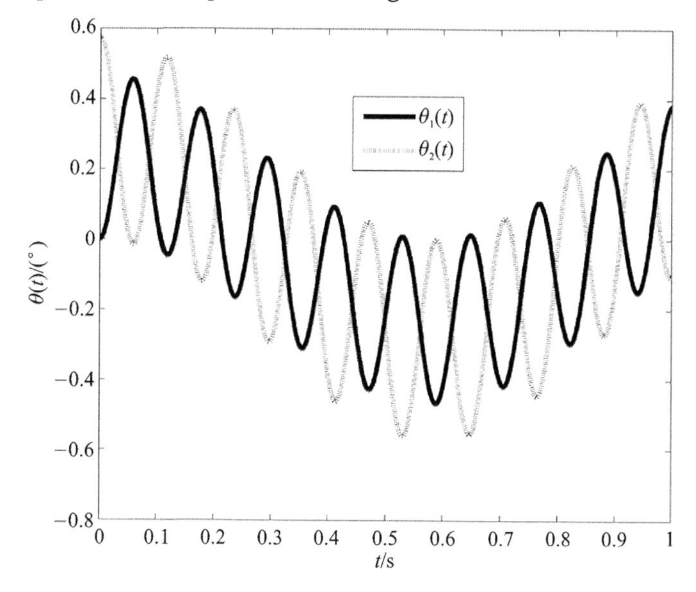

Fig. 5.25 The time response for Example 5.10 (g = 9.8 m/s in this case)

5.6.2 Harmonic excitation of undamped systems

If an undamped MDOF system is excited by the force $\boldsymbol{f}(t) = \boldsymbol{F}\sin(\omega t)$, $\boldsymbol{F}$ is a constant vector, and ω is the excitation frequency, then the matrix differential equation governing the vibrations is

$$\boldsymbol{M}\ddot{\boldsymbol{x}} + \boldsymbol{K}\boldsymbol{x} = \boldsymbol{F}\sin(\omega t) \tag{5.100}$$

We try a particular solution of the form

$$\boldsymbol{x}_{\mathrm{p}} = \boldsymbol{u}\sin(\omega t) \tag{5.101}$$

Substituting Eq. (5.101) into Eq. (5.100), we get

$$(\boldsymbol{K} - \omega^2\boldsymbol{M})\boldsymbol{u} = \boldsymbol{F} \tag{5.102}$$

So if $(\boldsymbol{K} - \omega^2\boldsymbol{M})$ is invertible, we get

$$\boldsymbol{u} = (\boldsymbol{K} - \omega^2\boldsymbol{M})^{-1}\boldsymbol{F} \tag{5.103}$$

According to the discussion in Subsection 5.7.1, if ω is equal to one of the natural frequencies, the matrix $(\boldsymbol{K} - \omega^2\boldsymbol{M})$ is noninvertible and singular. In the case there is generally no vector $\boldsymbol{u}$ which satisfies Eq. (5.102), because the system is at resonance and the vector $\boldsymbol{u}$ will tend to infinite when the exciting frequency ω is equal to one of natural frequencies of the system.

The general solution of Eq. (5.100) can be written as

$$\boldsymbol{x} = \boldsymbol{x}_{\mathrm{p}} + \boldsymbol{x}_{\mathrm{h}} \tag{5.104}$$

where $\boldsymbol{x}_{\mathrm{p}}$ is the particular solution shown in Eq. (101), $\boldsymbol{x}_{\mathrm{h}}$ is the homogeneous solution of the homogeneous problem shown in Eqs. (5.86) and (5.93).

So the general solution of Eq. (5.100) can be written as

$$\boldsymbol{x}(t) = \begin{cases} \sum_{i=1}^{n} \left[a_i\boldsymbol{X}_i\sin(\sqrt{\lambda_i}t) + b_i\boldsymbol{X}_i\cos(\sqrt{\lambda_i}t)\right] + \boldsymbol{u}\sin(\omega t) & \lambda_1 \neq 0 \\ a_1 t\boldsymbol{X}_1 + b_1\boldsymbol{X}_1 + \sum_{i=2}^{n} \left[a_i\boldsymbol{X}_i\sin(\sqrt{\lambda_i}t) + b_i\boldsymbol{X}_i\cos(\sqrt{\lambda_i}t)\right] + \boldsymbol{u}\sin(\omega t) & \lambda_1 = 0 \end{cases} \tag{5.105}$$

Example 5.11 (Example 5.8 revisited)

In Example 5.8, if a applied force $f(t)$ is added, as shown in Fig. 5.26, assume that the system is with zero initial conditions. Determine the response of the same system.

Solution:

The differential equation of motion is

$$\boldsymbol{M}\ddot{\boldsymbol{x}} + \boldsymbol{K}\boldsymbol{x} = \boldsymbol{f}(t)$$

with

$$\boldsymbol{M} = \begin{bmatrix} 1 & 0 \\ 0 & 1 \end{bmatrix}, \quad \boldsymbol{K} = \begin{bmatrix} 3 & -2 \\ -2 & 3 \end{bmatrix}, \quad \boldsymbol{f}(t) = \begin{bmatrix} 0 \\ 1 \end{bmatrix} \sin 2t$$

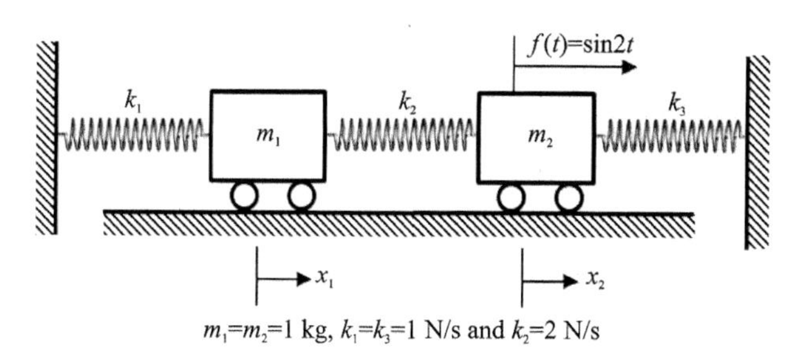

Fig. 5.26 Forced vibration of a two-degree-of-freedom system

Notice that there are zero initial conditions, and according to Eq. (5.105), the general solution in this case is

$$x = \boldsymbol{u}\sin(\omega t) \quad \text{and} \quad \boldsymbol{u} = (\boldsymbol{K} - \omega^2\boldsymbol{M})^{-1}\boldsymbol{f}$$

It can easily be solved using MATLAB (also it is possible to solve it by hand). For example, we can input the commands in MATLAB:

```
M=[1 0; 0 1];
K=[3 -2; -2 3];
omega=2;
f = [0; 1];
u =inv(K-M * omega^2) * f
```

We can get vector $\boldsymbol{u}$ immediately, that is

$$\boldsymbol{u} = \begin{bmatrix} -0.667 \\ 0.333 \end{bmatrix}$$

So the solution for this example is

$$\boldsymbol{x} = \begin{bmatrix} -0.667 \\ 0.333 \end{bmatrix} \sin 2t$$

If the excitation force in Fig. 5.26 is changed as $f(t) = \sin(\omega t)$. Fig. 5.27 shows the vibration amplitude $|\boldsymbol{u}|$ with different excitation frequencies. Notice that at some frequencies $\boldsymbol{x}$ has negative vibration amplitudes, as the negative sign just means that the mass vibrates out of phase with the applied force. So we plot the absolute results of $\boldsymbol{x}$.

From Fig. 5.27, several important features can be found:

(1) Notice that the natural frequencies of the system are 1 rad/s and 2.236 rad/s, as presented in Example 5.8. If the forcing frequency is close to any one of the natural frequencies of the system, large vibration amplitudes occur. This phenomenon is known as **resonance**. At these frequencies the vibration amplitude is theoretically infinite.

(2) Another important phenomenon can be found in Fig. 5.27. That is, the

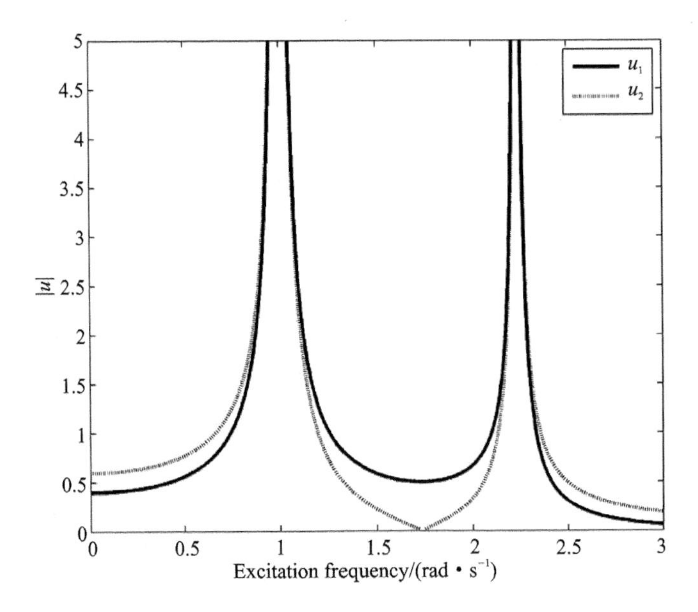

Fig. 5.27 The vibration amplitude $|u|$ with different excitation frequencies

amplitude of vibration of Mass 2 (that the force acts on) drops to zero at one special excitation frequency. This is called "**anti-resonance**", and it has an important engineering application, such as vibration control or vibration absorbing.

5.7 Vibration of Underdamped MDOF Systems

In this section we deal with the damped vibration problem for MDOF system briefly.

5.7.1 Free vibration of underdamped MDOF systems

Fig. 5.28 shows a spring-mass-damper system, for free vibration case, the excitation forces $F_1(t)=F_2(t) = 0$. The equations of motion for the system can easily be obtained as

$$\begin{bmatrix} m_1 & 0 \\ 0 & m_2 \end{bmatrix}\begin{bmatrix} \ddot{x}_1 \\ \ddot{x}_2 \end{bmatrix}+\begin{bmatrix} c_1+c_2 & -c_2 \\ -c_2 & c_2+c_3 \end{bmatrix}\begin{bmatrix} \dot{x}_1 \\ \dot{x}_2 \end{bmatrix}+\begin{bmatrix} k_1+k_2 & -k_2 \\ -k_2 & k_2+k_3 \end{bmatrix}\begin{bmatrix} x_1 \\ x_2 \end{bmatrix}=\begin{bmatrix} 0 \\ 0 \end{bmatrix} \tag{5.106}$$

To solve Eq. (5.106), we have to modify them to a system that MATLAB can handle, by rewriting them as first order equations. We follow the standard procedure to do this, define

$$\begin{bmatrix} v_1 \\ v_2 \end{bmatrix}=\begin{bmatrix} \dot{x}_1 \\ \dot{x}_2 \end{bmatrix} \tag{5.107a}$$

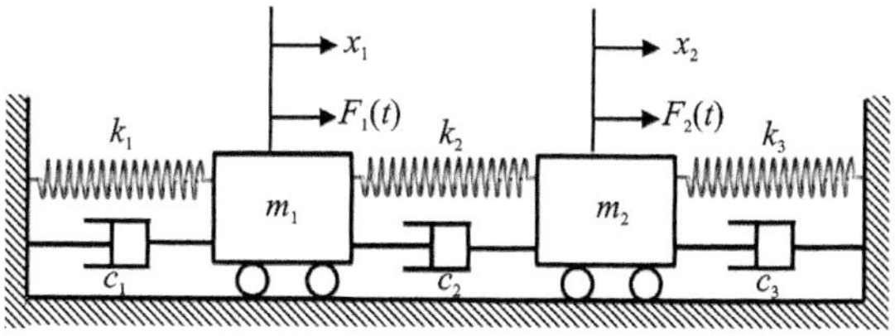

Fig. 5.28 A damped spring-mass system

Thus

$$\begin{bmatrix} \dot{v}_1 \\ \dot{v}_2 \end{bmatrix} = \begin{bmatrix} \ddot{x}_1 \\ \ddot{x}_2 \end{bmatrix} \tag{5.107b}$$

then Eq. (5.106) can be rewritten as

$$\begin{bmatrix} 1 & 0 & 0 & 0 \\ 0 & 1 & 0 & 0 \\ 0 & 0 & m_1 & 0 \\ 0 & 0 & 0 & m_2 \end{bmatrix} \begin{bmatrix} \dot{x}_1 \\ \dot{x}_2 \\ \dot{v}_1 \\ \dot{v}_2 \end{bmatrix} + \begin{bmatrix} 0 & 0 & -1 & 0 \\ 0 & 0 & 0 & -1 \\ k_1+k_2 & -k_2 & c_1+c_2 & -c_2 \\ -k_2 & k_2+k_3 & -c_2 & c_2+c_3 \end{bmatrix} \begin{bmatrix} x_1 \\ x_2 \\ v_1 \\ v_2 \end{bmatrix} = \begin{bmatrix} 0 \\ 0 \\ 0 \\ 0 \end{bmatrix} \tag{5.108}$$

Eq. (5.108) is a matrix equation of the form:

$$\boldsymbol{Q}\dot{\boldsymbol{y}} + \boldsymbol{D}\boldsymbol{y} = \boldsymbol{0} \tag{5.109}$$

where

$$\boldsymbol{Q} = \begin{bmatrix} 1 & 0 & 0 & 0 \\ 0 & 1 & 0 & 0 \\ 0 & 0 & m_1 & 0 \\ 0 & 0 & 0 & m_2 \end{bmatrix}, \quad \boldsymbol{D} = \begin{bmatrix} 0 & 0 & -1 & 0 \\ 0 & 0 & 0 & -1 \\ k_1+k_2 & -k_2 & c_1+c_2 & -c_2 \\ -k_2 & k_2+k_3 & -c_2 & c_2+c_3 \end{bmatrix}, \quad \boldsymbol{y} = \begin{bmatrix} x_1 \\ x_2 \\ v_1 \\ v_2 \end{bmatrix}$$

Suppose that at time $t=0$, the system has initial conditions as follows:

$$\left.\begin{aligned} x(0) &= x_0 \\ \dot{x}(0) &= \boldsymbol{v}_0 \end{aligned}\right\} \tag{5.110}$$

We start by guessing that the solution has the form as follows:

$$\boldsymbol{y} = \boldsymbol{U}\exp(-\lambda t) \tag{5.111}$$

where $\boldsymbol{U}$ is a constant vector, to be determined. And the negative sign is introduced because we expect solutions to decay with time.

Substituting Eq. (5.111) into Eq. (5.109), we get

$$-\boldsymbol{Q}\boldsymbol{\lambda}\boldsymbol{U}\exp(-\lambda t) + \boldsymbol{D}\boldsymbol{U}\exp(-\lambda t) = \boldsymbol{0} \tag{5.112}$$

and

$$(\boldsymbol{D} - \lambda\boldsymbol{Q})\boldsymbol{U} = \boldsymbol{0} \tag{5.113}$$

Notice that Eq. (5.113) has the same formula to Eq. (5.84), and this is another generalized eigenvalue problem, and can easily be solved with MATLAB.

However, it should be noted that the solution is much more complicated for a damped system, because the possible values of $\boldsymbol{U}$ and λ that satisfy Eq. (5.113) are in general complex. Based on the Euler's formula, it can be found that each λ can be expressed as

$$\lambda = \zeta \pm \mathrm{i}\omega \quad \text{and} \quad \mathrm{i} = \sqrt{-1} \tag{5.114}$$

where ζ and ω are the positive real numbers.

Similar to Subsection 5.7.1, based on Eq. (5.111), it is easy to find that solution to Eq. (5.109) can be expressed as

$$y(t) = \sum_{i=1}^{2n} [a_i \boldsymbol{U}_i \exp(-\lambda_i t)] \tag{5.115}$$

where a_i is unknown coefficients, which depend on the initial conditions.

Clearly, Eq. (5.115) satisfies the differential Eq. (5.109). But $y(t)$ in Eq. (5.115) is expressed in terms of $2n$ arbitrary constants and it is the general solution of Eq. (5.109).

The coefficients a_i can be determined by the initial conditions in Eq. (5.110). Using Eqs. (5.110) and (5.115), the initial condition can be rewritten as

$$\boldsymbol{y}_0 = \begin{bmatrix} \boldsymbol{x}_0 \\ \boldsymbol{v}_0 \end{bmatrix} = \sum_{i=1}^{n} [a_i \boldsymbol{U}_i] = \boldsymbol{U}\boldsymbol{a} \tag{5.116}$$

where matrix $\boldsymbol{U} = [\boldsymbol{U}_1, \boldsymbol{U}_2, \cdots, \boldsymbol{U}_n]$, and $\boldsymbol{a} = [a_1, a_2, \cdots, a_n]^{\mathrm{T}}$.

We thus have

$$\boldsymbol{a} = \boldsymbol{U}^{-1} \boldsymbol{y}_0 \tag{5.117}$$

The following example demonstrates this result.

Example 5.12

Assume that the damped spring-mass system shown in Fig. 5.28 has parameters $m_1 = m_2 = 1$, $k_1 = k_2 = k_3 = 1$ and $c_1 = c_2 = c_3 = 0.5$. Assume that the initial conditions of the system are

$$\begin{bmatrix} x_1(0) \\ x_2(0) \end{bmatrix} = \begin{bmatrix} 1 \\ 1 \end{bmatrix}, \quad \begin{bmatrix} \dot{x}_1(0) \\ \dot{x}_2(0) \end{bmatrix} = \begin{bmatrix} 0 \\ 0 \end{bmatrix}$$

Determine the response of the system when the excitation forces $F_1(t) = F_2(t) = 0$.

Solution:

From Eq. (5.109), the equations of motion for the system

$$\boldsymbol{Q}\dot{\boldsymbol{y}} + \boldsymbol{D}\boldsymbol{y} = \boldsymbol{0}$$

with

$$\boldsymbol{Q} = \begin{bmatrix} 1 & 0 & 0 & 0 \\ 0 & 1 & 0 & 0 \\ 0 & 0 & 1 & 0 \\ 0 & 0 & 0 & 1 \end{bmatrix}, \quad \boldsymbol{D} = \begin{bmatrix} 0 & 0 & -1 & 0 \\ 0 & 0 & 0 & -1 \\ 2 & -1 & 1 & -0.5 \\ -1 & 2 & -0.5 & 1 \end{bmatrix}, \quad \boldsymbol{y} = \begin{bmatrix} x_1 \\ x_2 \\ v_1 \\ v_2 \end{bmatrix}$$

By using MATLAB function eig, as shown in Section 5.6, the $\boldsymbol{\Lambda}$ (with diagonal element λ_i) and $\boldsymbol{U}$ matrices for this system are

$$\boldsymbol{\Lambda}=\begin{bmatrix}0.75+1.561i & 0 & 0 & 0\\ 0 & 0.75-1.561i & 0 & 0\\ 0 & 0 & 0.25+0.968i & 0\\ 0 & 0 & 0 & 0.25-0.968i\end{bmatrix}$$

$$\boldsymbol{U}=\begin{bmatrix}-0.315+0.335i & -0.315-0.335i & -0.434+0.566i & -0.434-0.566i\\ 0.315-0.335i & 0.315+0.335i & -0.434+0.566i & -0.434-0.566i\\ 0.759+0.241i & 0.759-0.241i & 0.656+0.279i & 0.656-0.279i\\ -0.759-0.241i & -0.759+0.241i & 0.656+0.279i & 0.656-0.279i\end{bmatrix}$$

By using Eq. (5.117) and initial conditions, we get

$$\boldsymbol{a}=\boldsymbol{U}^{-1}\boldsymbol{y}_0=\begin{bmatrix}0\\ 0\\ -0.283-0.666i\\ -0.283+0.666i\end{bmatrix}$$

So the general solution for this problem can be obtained by using Eq. (5.115), such as

$$\boldsymbol{y}(t)=\begin{bmatrix}x(t)\\ \dot{x}(t)\end{bmatrix}=\sum_{i=1}^{2n}[a_i\boldsymbol{U}_i\exp(-\lambda_i t)]$$

$$=(-0.283-0.666i)\begin{bmatrix}-0.434+0.566i\\ -0.434+0.566i\\ 0.656+0.279i\\ 0.656+0.279i\end{bmatrix}\exp[(0.25+0.968i)t]+$$

$$(-0.283+0.666i)\begin{bmatrix}-0.434-0.566i\\ -0.434-0.566i\\ 0.656-0.279i\\ 0.656-0.279i\end{bmatrix}\exp[(0.25-0.968i)t]$$

Clearly, the displacement response of the system is

$$\boldsymbol{x}(t)=\begin{bmatrix}x_1(t)\\ x_2(t)\end{bmatrix}=(-0.283-0.666i)\begin{bmatrix}-0.434+0.566i\\ -0.434+0.566i\end{bmatrix}\exp[(0.25+0.968i)t]+$$

$$(-0.283+0.666i)\begin{bmatrix}-0.434-0.566i\\ -0.434-0.566i\end{bmatrix}\exp[(0.25-0.968i)t]$$

We note that, all the matrices and vectors in these formulas are complex values. However, all the imaginary parts will disappear in the final answer.

Fig. 5.29 shows the response of this example. From Fig. 5.29, it can be found that two masses have the exact same displacements due to the initial conditions.

Changed the initial conditions, the responses will be changed, as shown in

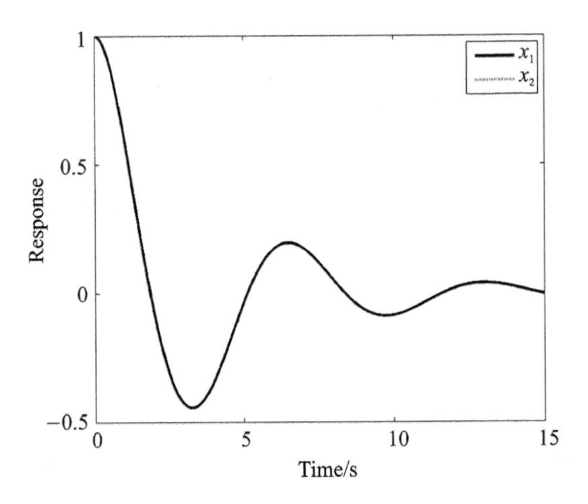

Fig. 5.29 Response for a damped spring-mass system

Fig. 5.30.

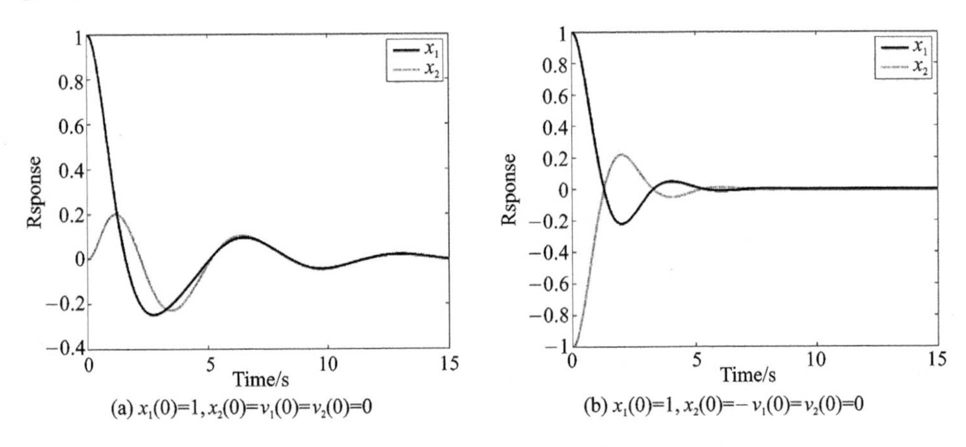

(a) $x_1(0)=1, x_2(0)=v_1(0)=v_2(0)=0$

(b) $x_1(0)=1, x_2(0)=-v_1(0)=v_2(0)=0$

Fig. 5.30 Response for a damped spring-mass system

5.7.2 Steady-state forced vibration response for MDOF systems

Finally, we take a look at the effects of applied forces on the response of a spring-mass system to harmonic forces with $f_1(t)=F_1 \cdot \cos(\omega t)$ and $f_2(t)=F_2 \cdot \cos(\omega t)$, as shown in Fig. 5.28. From Eq. (5.108), the equations of motion for a damped and forced system can be expressed as

$$\begin{bmatrix} 1 & 0 & 0 & 0 \\ 0 & 1 & 0 & 0 \\ 0 & 0 & m_1 & 0 \\ 0 & 0 & 0 & m_2 \end{bmatrix} \begin{bmatrix} \dot{x}_1 \\ \dot{x}_2 \\ \dot{v}_1 \\ \dot{v}_2 \end{bmatrix} + \begin{bmatrix} 0 & 0 & -1 & 0 \\ 0 & 0 & 0 & -1 \\ k_1+k_2 & -k_2 & c_1+c_2 & -c_2 \\ -k_2 & k_2+k_3 & -c_2 & c_2+c_3 \end{bmatrix} \begin{bmatrix} x_1 \\ x_2 \\ v_1 \\ v_2 \end{bmatrix} = \begin{bmatrix} 0 \\ 0 \\ F_1 \\ F_2 \end{bmatrix} \cos(\omega t) \tag{5.118}$$

By using Euler's formula, we get

$$\cos(\omega t) = \frac{\exp(\mathrm{i}\omega t) + \exp(-\mathrm{i}\omega t)}{2} \tag{5.119}$$

Similar to the discussion in Subsection 5.8.1, Eq. (5.118) can be expressed as a matrix equation of the form, such as

$$\boldsymbol{Q}\dot{\boldsymbol{y}} + \boldsymbol{D}\boldsymbol{y} = \boldsymbol{f}\cos(\omega t) = \boldsymbol{f}\,\frac{\exp(\mathrm{i}\omega t) + \exp(-\mathrm{i}\omega t)}{2} \tag{5.120}$$

where

$$\boldsymbol{Q} = \begin{bmatrix} 1 & 0 & 0 & 0 \\ 0 & 1 & 0 & 0 \\ 0 & 0 & m_1 & 0 \\ 0 & 0 & 0 & m_2 \end{bmatrix}, \quad \boldsymbol{D} = \begin{bmatrix} 0 & 0 & -1 & 0 \\ 0 & 0 & 0 & -1 \\ k_1 + k_2 & -k_2 & c_1 + c_2 & -c_2 \\ -k_2 & k_2 + k_3 & -c_2 & c_2 + c_3 \end{bmatrix},$$

$$\boldsymbol{y} = \begin{bmatrix} x_1 \\ x_2 \\ v_1 \\ v_2 \end{bmatrix}, \quad \boldsymbol{f} = \begin{bmatrix} 0 \\ 0 \\ F_1 \\ F_2 \end{bmatrix}$$

Here, we assume that the solution for Eq. (5.120) is

$$\boldsymbol{y} = \boldsymbol{y}_1 + \boldsymbol{y}_2 = \boldsymbol{U}_1\exp(\mathrm{i}\omega t) + \boldsymbol{U}_2\exp(-\mathrm{i}\omega t) \tag{5.121}$$

where $\boldsymbol{y}_1 = \boldsymbol{U}_1\exp(\mathrm{i}\omega t)$, $\boldsymbol{y}_2 = \boldsymbol{U}_2\exp(-\mathrm{i}\omega t)$.

Substituting Eq. (5.121) into Eq. (5.120), we get

$$\boldsymbol{Q}\dot{\boldsymbol{y}} + \boldsymbol{D}\boldsymbol{y} = \boldsymbol{Q}(\dot{\boldsymbol{y}}_1 + \dot{\boldsymbol{y}}_2) + \boldsymbol{D}(\boldsymbol{y}_1 + \boldsymbol{y}_2) = \frac{\boldsymbol{f}}{2}\exp(\mathrm{i}\omega t) + \frac{\boldsymbol{f}}{2}\exp(-\mathrm{i}\omega t) \tag{5.122}$$

and Eq. (5.122) can be modified as

$$\boldsymbol{Q}\dot{\boldsymbol{y}}_1 + \boldsymbol{D}\boldsymbol{y}_1 = \frac{\boldsymbol{f}}{2}\exp(\mathrm{i}\omega t) \tag{5.123a}$$

$$\boldsymbol{Q}\dot{\boldsymbol{y}}_2 + \boldsymbol{D}\boldsymbol{y}_2 = \frac{\boldsymbol{f}}{2}\exp(-\mathrm{i}\omega t) \tag{5.123b}$$

Substituting $\boldsymbol{y}_1 = \boldsymbol{U}_1\exp(\mathrm{i}\omega t)$ and $\boldsymbol{y}_2 = \boldsymbol{U}_2\exp(-\mathrm{i}\omega t)$ into Eqs. (5.123a) and (5.123b), respectively, we get

$$(\mathrm{i}\omega\boldsymbol{Q} + \boldsymbol{D})\boldsymbol{U}_1\exp(\mathrm{i}\omega t) = \frac{\boldsymbol{f}}{2}\exp(\mathrm{i}\omega t) \tag{5.124a}$$

$$(-\mathrm{i}\omega\boldsymbol{Q} + \boldsymbol{D})\boldsymbol{U}_2\exp(-\mathrm{i}\omega t) = \frac{\boldsymbol{f}}{2}\exp(-\mathrm{i}\omega t) \tag{5.124b}$$

From Eq. (5.124), we get

$$(\mathrm{i}\omega\boldsymbol{Q} + \boldsymbol{D})\boldsymbol{U}_1 = \frac{\boldsymbol{f}}{2} \quad \text{and} \quad \boldsymbol{U}_1 = (\mathrm{i}\omega\boldsymbol{Q} + \boldsymbol{D})^{-1}\,\frac{\boldsymbol{f}}{2} \tag{5.125a}$$

$$(-\mathrm{i}\omega \boldsymbol{Q} + \boldsymbol{D})\boldsymbol{U}_2 = \frac{\boldsymbol{f}}{2} \quad \text{and} \quad \boldsymbol{U}_2 = (-\mathrm{i}\omega \boldsymbol{Q} + \boldsymbol{D})^{-2} \frac{\boldsymbol{f}}{2} \qquad (5.125\text{b})$$

Substituting the solved $\boldsymbol{U}_1$ and $\boldsymbol{U}_2$ into Eq. (5.121), the full solution is obtained. Notice that there is no initial condition involved, it is the steady-state vibration response.

Given the force vector $\boldsymbol{f}$, excitation frequency ω, the matrices $\boldsymbol{Q}$ and $\boldsymbol{D}$ that describe the system, Eq. (5.125) can be simply solved by MATLAB as follows:

```
i=sqrt(-1);
U1 = inv(D+Q * i * omega) * f/2;
U2 = inv(D-Q * i * omega) * f/2;
```

Example 5.13

Determine the steady-state response of the system shown in Fig. 5.28. Assume that $m_1 = m_2 = 1$, $k_1 = k_2 = k_3 = 1$ and $c_1 = c_2 = c_3 = 0.5$. The excitation forces is $F_1(t) = 0$, and $F_2(t) = \cos 2t$.

Solution:

From Eq. (5.118), the equations of motion for the system

$$\boldsymbol{Q}\dot{\boldsymbol{y}} + \boldsymbol{D}\boldsymbol{y} = \boldsymbol{f}\cos 2t$$

with

$$\boldsymbol{Q} = \begin{bmatrix} 1 & 0 & 0 & 0 \\ 0 & 1 & 0 & 0 \\ 0 & 0 & 1 & 0 \\ 0 & 0 & 0 & 1 \end{bmatrix}, \quad \boldsymbol{D} = \begin{bmatrix} 0 & 0 & -1 & 0 \\ 0 & 0 & 0 & -1 \\ 2 & -1 & 1 & -0.5 \\ -1 & 2 & -0.5 & 1 \end{bmatrix}, \quad \boldsymbol{y} = \begin{bmatrix} x_1 \\ x_2 \\ v_1 \\ v_2 \end{bmatrix}, \quad \boldsymbol{f} = \begin{bmatrix} 0 \\ 0 \\ 0 \\ 1 \end{bmatrix}$$

by using Eq. (5.125), we get

$$\boldsymbol{U}_1 = (\mathrm{i}\omega \boldsymbol{Q} + \boldsymbol{D})^{-1} \frac{\boldsymbol{f}}{2} = \begin{bmatrix} -0.05 + 0.05\mathrm{i} \\ -0.1 - 0.1\mathrm{i} \\ -0.1 - 0.1\mathrm{i} \\ 0.2 - 0.2\mathrm{i} \end{bmatrix}$$

$$\boldsymbol{U}_2 = (-\mathrm{i}\omega \boldsymbol{Q} + \boldsymbol{D})^{-2} \frac{\boldsymbol{f}}{2} = \begin{bmatrix} -0.05 - 0.05\mathrm{i} \\ -0.1 + 0.1\mathrm{i} \\ -0.1 + 0.1\mathrm{i} \\ 0.2 + 0.2\mathrm{i} \end{bmatrix}$$

Base on Eq. (5.121), the solution can be obtained as follows:

$$\boldsymbol{y} = \begin{bmatrix} x_1(t) \\ x_2(t) \\ v_1(t) \\ v_2(t) \end{bmatrix} = \begin{bmatrix} -0.05 + 0.05\mathrm{i} \\ -0.1 - 0.1\mathrm{i} \\ -0.1 - 0.1\mathrm{i} \\ 0.2 - 0.2\mathrm{i} \end{bmatrix} \exp(2\mathrm{i}t) + \begin{bmatrix} -0.05 - 0.05\mathrm{i} \\ -0.1 + 0.1\mathrm{i} \\ -0.1 + 0.1\mathrm{i} \\ 0.2 + 0.2\mathrm{i} \end{bmatrix} \exp(-2\mathrm{i}t)$$

Figs. 5.31 and 5.32 show the vibration amplitude under different damping.

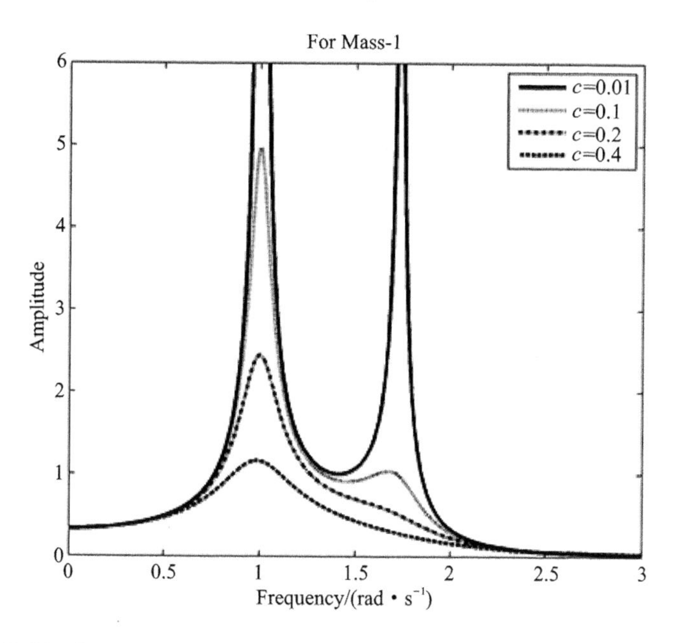

Fig. 5.31 The vibration amplitude for Mass-1 against the excitation frequency

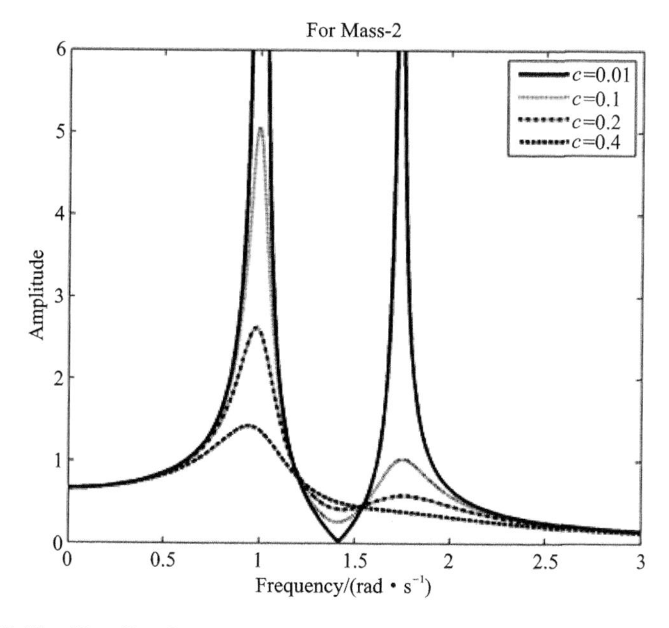

Fig. 5.32 The vibration amplitude for Mass-2 against the excitation frequency

Note that only Mass-2 is subjected to a force. From this example, some useful conclusions can be found:

(1) Resonance will occur when the driving frequency matches one of the

natural frequencies of the system. There are two resonances, at frequencies very close to the undamped natural frequencies of the system.

(2) In a damped system, it can be deduced that the higher the natural frequency, the lower the amplitude of vibration. Hence the amplitude of the lowest frequency resonance is generally much larger than higher frequency resonances. For this reason, it is often sufficient to consider only the first several natural frequencies and corresponding mode shapes in design calculations. Although it is possible to choose a set of forces that will excite only a high frequency mode, in which case the amplitude of this special excited mode will larger than the others. But for most applied forces, the lowest natural frequency is generally the most important one.

(3) In Fig. 5.32, it can be found that the "anti-resonance" behavior shown by the forced mass disappears as the damping increases.

5.8 The Dynamic Vibration Absorber

From analysis of the damped vibration system, it can be found that the adding damping can reduce the vibration. And there is a wide range of damping devices commercially available. However, if there is a particularly troublesome resonance, it may be preferable to add a vibration absorber. This is simply a spring-mass system which is added to the structure; the parameters of the absorber are chosen so that the amplitude of the vibration of the structure is greatly reduced or even eliminated at the targeted frequency.

If a SDOF system or one mode of a MDOF system is excited into resonance, very large amplitudes of vibration result with accompanying high dynamic stresses and noise and fatigue problems. In most mechanical systems this is not acceptable.

If neither the excitation frequency nor the natural frequency can conveniently be altered, this resonance condition can often be successfully controlled by adding a further SDOF system as vibration absorber.

Consider the model of the system shown in Fig. 5.33 where k_p and m_p are the effective stiffness and mass of the primary system when vibrating in the troublesome mode. And it is excited by harmonic force $f(t) = F_0 \sin(\omega t)$, as shown in Fig. 5.33.

Our objective is to eliminate the component of the forced vibration of the primary mass m_p. This is achieved by attaching a secondary SDOF system (vibration absorber), with mass m_s and spring k_s, to the primary system. The problem is to choose the right parameters, m_s and k_s, of the vibration absorber.

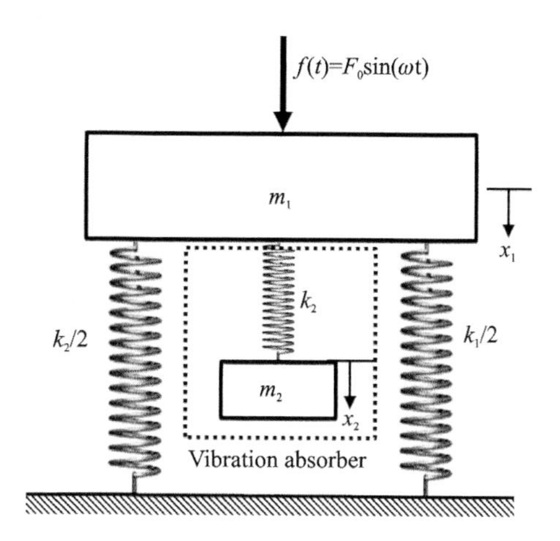

Fig. 5.33 System with undamped vibration absorber

From Subsection 5.1.1, it can be seen that the equations of motion are

$$\begin{bmatrix} m_p & 0 \\ 0 & m_s \end{bmatrix} \begin{bmatrix} \ddot{x}_p \\ \ddot{x}_s \end{bmatrix} + \begin{bmatrix} k_p + k_s & -k_s \\ -k_s & k_s \end{bmatrix} \begin{bmatrix} x_p \\ x_s \end{bmatrix} = \begin{bmatrix} F_0 \\ 0 \end{bmatrix} \sin(\omega t) \tag{5.126}$$

The forced harmonic vibrations of the undamped system have solution of the form

$$\begin{bmatrix} x_p \\ x_s \end{bmatrix} = \begin{bmatrix} X_p \\ X_s \end{bmatrix} \sin(\omega t) \tag{5.127}$$

Substituting Eq. (5.127) into Eq. (5.126), we get

$$\begin{bmatrix} -\omega^2 m_p + (k_p + k_s) & -k_s \\ -k_s & -\omega^2 m_s + k_s \end{bmatrix} \begin{bmatrix} x_p \\ x_s \end{bmatrix} \sin(\omega t) = \begin{bmatrix} F_0 \\ 0 \end{bmatrix} \sin(\omega t) \tag{5.128}$$

In Eq. (5.128), the term $\sin(\omega t)$ can be eliminated, we get

$$\begin{bmatrix} -\omega^2 m_p + (k_p + k_s) & -k_s \\ -k_s & -\omega^2 m_s + k_s \end{bmatrix} \begin{bmatrix} x_p \\ x_s \end{bmatrix} = \begin{bmatrix} F_0 \\ 0 \end{bmatrix} \tag{5.129}$$

Notice that our goal is to eliminate the vibration of mass m_p, so we can set $x_p = 0$. And Eq. (5.129) is rewritten in component form

$$-k_s x_s = F_0 \tag{5.130a}$$

$$(-\omega^2 m_s + k_s) x_s = 0 \tag{5.130b}$$

From the Eq. (5.130b), it can be found that

$$\omega_o^2 = \frac{k_s}{m_s} \tag{5.131}$$

From Eq. (5.131), it can be found that if we choose the mass and stiffness of the absorber such that $k_s/m_s = \omega_o^2$, the vibration of the primary mass m_p will vanish at frequency ω_o.

The physical interpretation of this result is that: From Eq. (5.130a), we can find that the absorber applies a force to the primary system which is equal and opposite to the exciting force. Hence the mass in the primary system has a net zero exciting force acting on it and therefore zero vibration amplitude.

From above analysis, it can be found that if $\omega_o^2 = k_s/m_s = k_p/m_p$, then the response of the primary system at its original resonance frequency can be made zero. This is the usual tuning arrangement for an undamped vibration absorber because the resonance problem in the primary system is only severe when the excitation frequency ω is equal to the natural frequency ω_o.

If an absorber is correctly tuned, such as $\omega_o^2 = k_s/m_s = k_p/m_p$, and if the mass ratio $\mu = m_s/m_p$, then Eq. (5.129) can be rewritten as

$$\begin{bmatrix} -\left(\frac{\omega}{\omega_o}\right)^2 + (1+\mu) & -\mu \\ -\mu & -\mu\left(\frac{\omega}{\omega_o}\right)^2 + \mu \end{bmatrix} \begin{bmatrix} x_p \\ x_s \end{bmatrix} = \frac{1}{k_p} \begin{bmatrix} F_0 \\ 0 \end{bmatrix} \tag{5.132}$$

From Eq. (5.132), the frequency equation is

$$\det\left(\begin{bmatrix} -\left(\frac{\omega}{\omega_o}\right)^2 + (1+\mu) & -\mu \\ -\mu & -\mu\left(\frac{\omega}{\omega_o}\right)^2 + \mu \end{bmatrix}\right) = \left(\frac{\omega}{\omega_o}\right)^4 - (2+\mu)\left(\frac{\omega}{\omega_o}\right)^2 + 1 = 0 \tag{5.133}$$

This is a quadratic equation in terms of $\left(\frac{\omega}{\omega_o}\right)^2$. Hence

$$\left(\frac{\omega}{\omega_o}\right)^2 = \left(1 + \frac{\mu}{2}\right)^2 - \sqrt{\mu + \frac{\mu^2}{4}} \tag{5.134a}$$

$$\left(\frac{\omega}{\omega_o}\right)^2 = \left(1 + \frac{\mu}{2}\right)^2 + \sqrt{\mu + \frac{\mu^2}{4}} \tag{5.134b}$$

So the natural frequencies with vibration absorber are

$$\omega_{1,2}^{\text{new}} = \omega_o \sqrt{\left(1 + \frac{\mu}{2}\right)^2 - \sqrt{\mu + \frac{\mu^2}{4}}} \tag{5.135a}$$

$$\omega_{1,2}^{\text{new}} = \omega_o \sqrt{\left(1 + \frac{\mu}{2}\right)^2 + \sqrt{\mu + \frac{\mu^2}{4}}} \tag{5.135b}$$

For a small μ, ω_1^{new}, and ω_2^{new} are very close to each other, and near to ω_o. If we increase the value of μ, gives better separation between ω_1^{new} and ω_2^{new} as shown in Fig. 5.34.

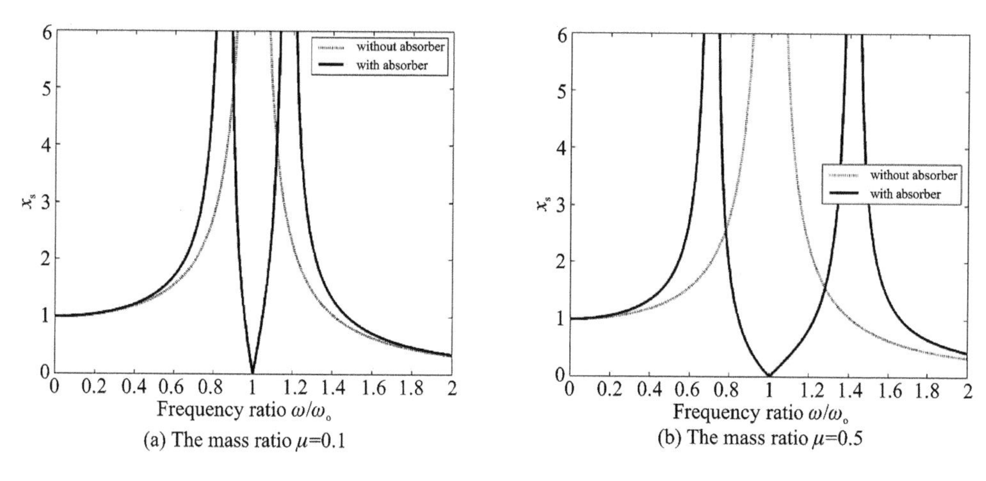

Fig. 5.34 Amplitude-frequency response for system with and without absorber

Questions

5.1 A two-degree-of-freedom system is shown in Fig. 5.35. Derive the equation of motion and calculate the mass and stiffness matrices.

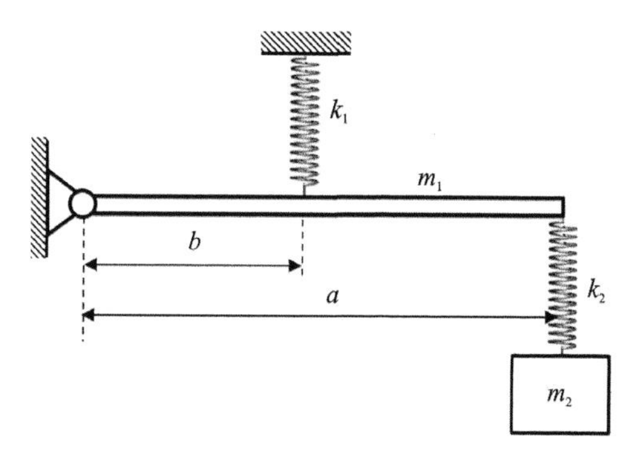

Fig. 5.35 A two-degree-of-freedom system

5.2 Consider the system of Fig. 5.36. Derive the equation of motion and calculate the mass and stiffness matrices.

5.3 Calculate the characteristic equation and solve for the system's natural frequencies from Question 5.2 (see Fig. 5.2).

(1) Suppose $m_1 = m$, $m_2 = 2m$ and $k_1 = k_2 = k_3 = k$;

(2) Suppose $m_1 = m_2 = 1$ kg, $k_1 = 1$ N/s, $k_2 = 10$ N/s and $k_3 = 100$ N/s.

5.4 Calculate the free response of the system of Question 5.3, assume that the initial conditions are $\boldsymbol{x}(0) = [2,4]^{\mathrm{T}}$ and $\boldsymbol{v}(0) = [0,0]^{\mathrm{T}}$.

5.5 Repeat Question 5.2 for the case that $k_1 = k_3 = 0$. And calculate and

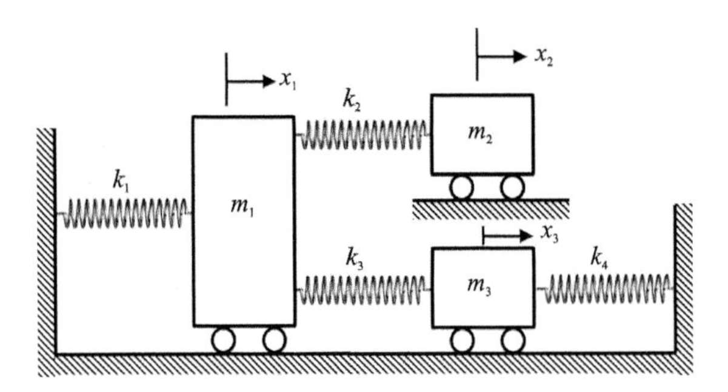

Fig. 5.36 A mass-spring system

solve the characteristic equation when $m_1 = 9$, $m_2 = 1$, $k_2 = 10$.

5.6 A torsional system is shown in Fig. 5.37. Determine the equation of motion in matrix form, then calculate the natural frequencies and mode shapes.

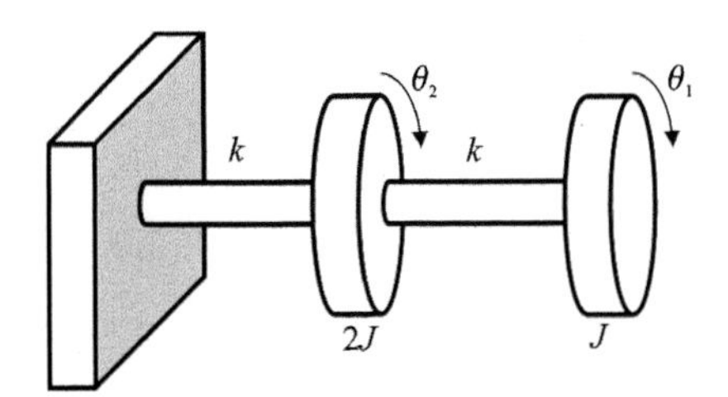

Fig. 5.37 A torsional system

5.7 A three-degree-of-freedom system is shown in Fig. 5.38. Determine the equation of motion in matrix form.

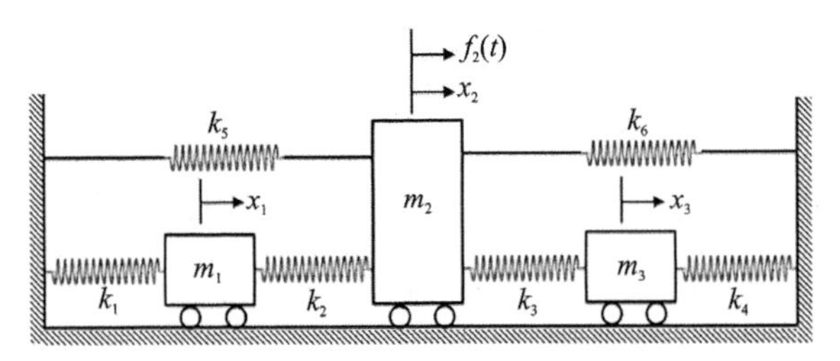

Fig. 5.38 A three-degree-of-freedom system

5.8 An undamped system is shown in Fig. 5.39.

(1) Determine the equation of motion in matrix form;

(2) Calculate the natural frequencies for this system, assume that $k_{R1} = k_{R2} = k_{R3} = k_{R4} = k$ and $J_1 = J_2/5 = J_3 = J$.

5.9 Consider the mass-spring system in Fig. 5.40. This system is free to vibration in the $x_1 - x_2$ plane. So each mass has two degrees of freedom.

(1) Derive the linearmotion equations, write them in matrix form;

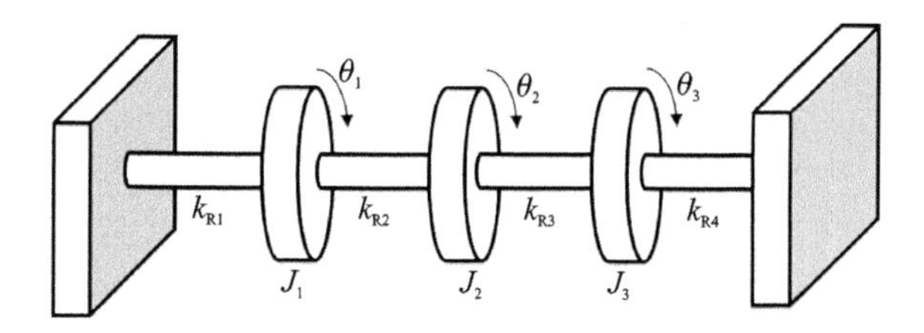

Fig. 5.39 An undamped system

(2) Assuming $k_1 = k$, $k_2 = 3k$, $k_3 = 6k$ and $k_4 = 3k$, calculate the natural frequencies and mode shapes.

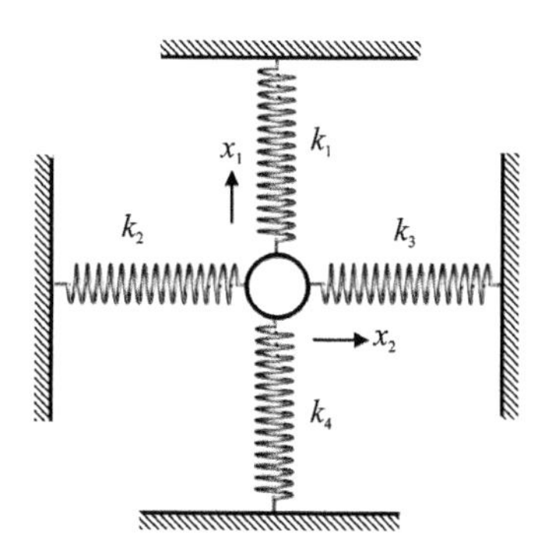

Fig 5.40 A mass-spring system vibration in the $x_1 - x_2$ plane

5.10 A car with mass vibrating on frictionless surface is shown in Fig. 5.41, assume that all surfaces are frictionless. Derive the linear equations of motion, write them in matrix form.

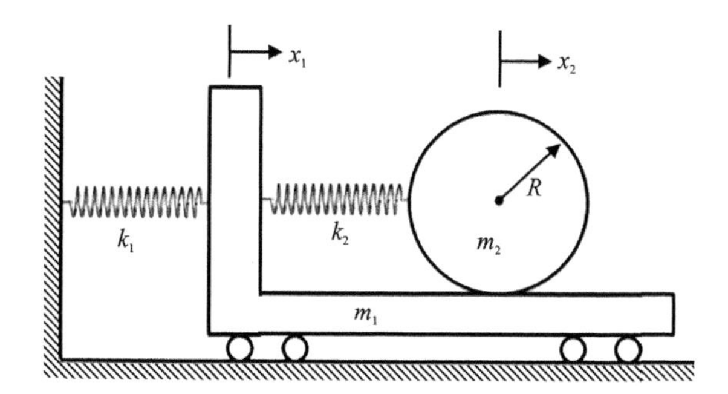

Fig. 5.41 A car with mass vibrating on frictionless surface

5.11 A trailer shown in Fig. 5.42 is connected by a coupler. The coupler can be modeled as a spring of stiffness k. Write the equation of motion and calculate the natural frequencies and mode shapes.

5.12 Consider the system shown in the Fig. 5.43, assume $m_1 = 4m_2$, $k_1 = 4k_2/5$. Write the equations of motion in matrix form and compute the natural

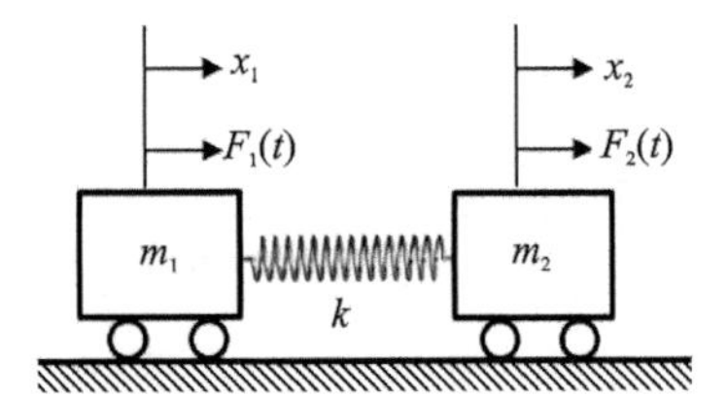

Fig. 5.42　A trailer on frictionless surface

frequencies and the mode shapes.

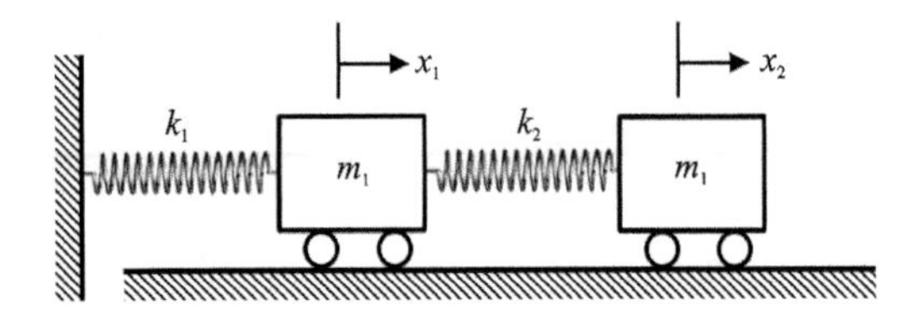

Fig. 5.43　A mass-spring system

5.13　Consider the system of Fig. 5.44, consisting of two pendulums coupled by a spring.

(1) Determine the natural frequencies;

(2) Calculate and plot the mode shapes.

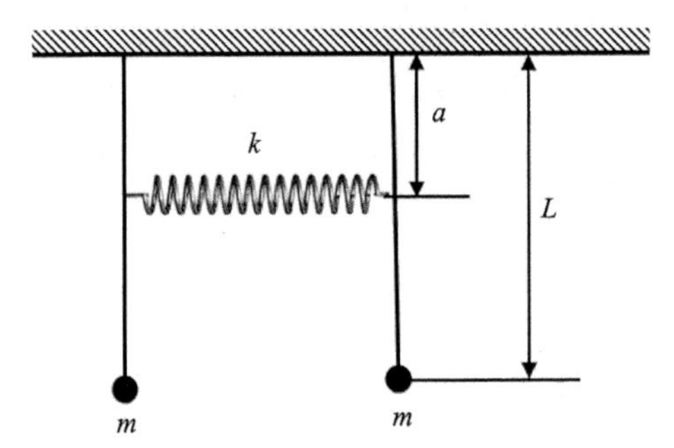

Fig. 5.44　A coupled two-pendulum system

5.14　Consider a vibration system shown in Fig. 5.45, a rigid bar with its gravity center not coinciding with its geometric center, and it is supported by two springs, k_1 and k_2. Write the equations of motion in matrix form.

5.15　Develop a MATLAB code for determination the natural frequencies and mode shapes of free-vibration for a general two-degree-of-freedom system.

5.16　Consider the two-degree-of-freedom system defined by

$$\boldsymbol{M}=\begin{bmatrix}1 & 0\\0 & 1\end{bmatrix},\quad \boldsymbol{K}=\begin{bmatrix}3 & -2\\-2 & 3\end{bmatrix}$$

Calculate the response of the system to the initial condition

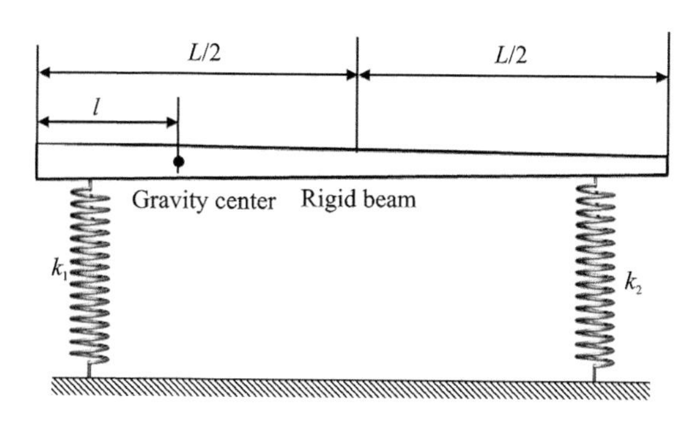

Fig. 5.45 A vibration model

$$\boldsymbol{x}_0 = \begin{bmatrix} 1 \\ 1 \end{bmatrix}, \quad \dot{\boldsymbol{x}}_0 = \begin{bmatrix} 0 \\ 0 \end{bmatrix}$$

5.17 Consider the system shown in Fig. 5.1. Let $k_1 = k_2 = k$ and $m_1 = m_2 = m$. Solve for the free response of this system with the initial conditions

$$\boldsymbol{x}_0 = \begin{bmatrix} 1 \\ -1 \end{bmatrix}, \quad \dot{\boldsymbol{x}}_0 = \begin{bmatrix} 0 \\ 0 \end{bmatrix}$$

5.18 The vibration is the vertical direction of an airplane and its wings can be modeled as a three-degree-of-freedom system as shown in Fig. 5.46. Masses m_1, m_2 and m_3 are corresponding to the left wing, the fuselage and left wing, respectively. Assume $L_1 = L_2$, $m_1 = m_3 = m$ and $m_2 = 5m$, and the bending stiffness of the wing is EI.

(1) Calculate the natural frequencies and mode shapes;

(2) Plot the mode shapes and interpret them according to the airplane's deflection.

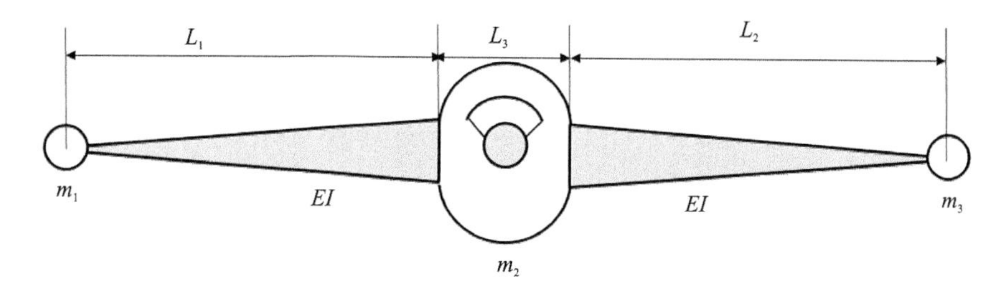

Fig. 5.46 An airplane modeled as a three-degree-of-freedom system

5.19 Consider the two-mass system of Fig. 5.47. This system is free to move in the $x_1 - x_2$ plane. Hence each mass has two degrees of freedom. Assume that $m_1 = m_2 = 1$ kg and $k = 100$ N/m.

(1) Derive the linear equations of motion and write them in matrix form;

(2) Calculate the natural frequencies;

(3) Calculate the responses for initial displacements $x_1(0) = 1$, $x_2(0) = x_3(0) = x_4(0) = 0$ and zero initial velocity.

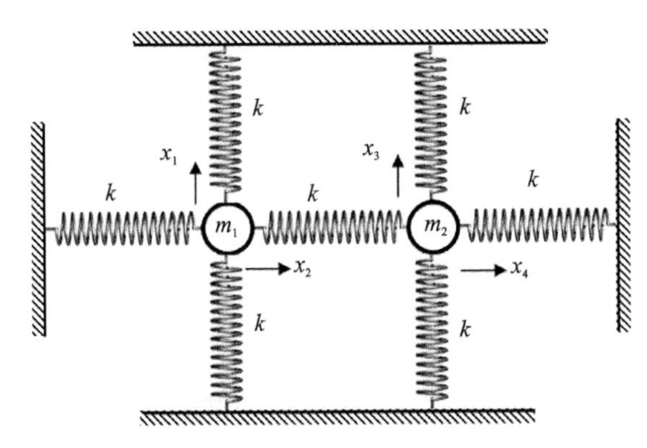

Fig. 5.47 A two-mass system

5.20 A damped mass-spring vibration system is shown in Fig. 5.48, The initial conditions are

$$\boldsymbol{x}_0 = \begin{bmatrix} 1 \\ 0 \end{bmatrix}, \quad \dot{\boldsymbol{x}}_0 = \begin{bmatrix} 0 \\ 0 \end{bmatrix}$$

(1) Derive the linear equations of motion, write them in matrix form.

(2) Calculate the free responses, suppose $m_1 = m$, $m_2 = 2m$, $k_1 = k_2 = k_3 = k$ and $c = 0.5$.

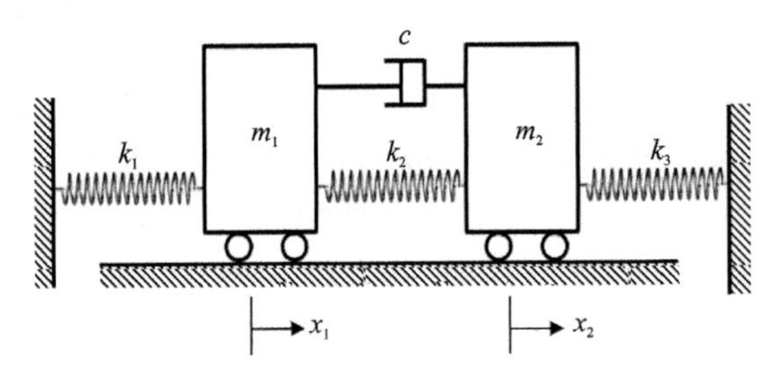

Fig. 5.48 A damped mass-spring system

Chapter 6 Vibration of Continuous Systems

Most of the vibration analysis examples encountered in the previous chapters assumed that the inertial (mass), flexibility (spring), and dissipative (damping) characteristics could be "lumped" as a finite number of "discrete" elements. Such models are termed lumped-parameter or discrete parameter systems.

Continuous structures such as beams, rods, cables and plates can be modelled by discrete mass and stiffness parameters and analyzed as MDOF systems, but such a model is not sufficiently accurate for most purposes. Furthermore, mass and elasticity cannot always be separated in models of real systems. Thus mass and elasticity have to be considered as distributed or continuous parameters. The present chapter concerns vibration analysis of continuous systems.

6.1 Transverse Vibration of Cables

The first continuous structure to be studied in this chapter is a cable in tension. This is a line structure for which its geometric configuration can be completely defined by the position of its axial line, with reference to a fixed coordinate line. We will study the transverse (lateral) vibration problem; that is, the vibration in a direction perpendicular to its axis and in a single plane.

The vibration problem of cables include stringed musical instruments, overhead transmission lines (of electric power or telephone signals), drive systems (belt drives, chain drives, pulley ropes, etc.), suspension bridges, and structural cables carrying cars (e.g., ski lifts, elevators, overhead sightseeing systems, and cable cars).

As usual, some simplifying assumptions will be made for analytical convenience; but the results and insight obtained in this manner will be useful in understanding the behavior of more complex systems containing cable-like structures. The main assumptions are

(1) The system is a line structure. The lateral dimensions are much smaller compared to the longitudinal dimension (normally in the x direction).

(2) The structure stays in a single plane, and the motion of every element of the structure will be in a fixed transverse direction (y).

(3) The cable tension (T) remains constant during motion. In other words, the initial tension is sufficiently large that the variations during motion are

negligible. It means that the tension T in the spring is independent of the time t and the position x.

(4) Variations in slope (θ) along the structure are small, so

$$\theta \simeq \sin\theta = \frac{\partial w}{\partial x} \tag{6.1}$$

Consider a uniform cable, as shown in Fig. 6.1. By using Newton's second law, the equation of motion (transverse) of one element of the cable shown in Fig. 6.1 can be expressed as

$$f(x,t)\mathrm{d}x - T\sin\theta + T\sin(\theta+\mathrm{d}\theta) = m(x)\mathrm{d}x\,\frac{\partial^2 w(x,t)}{\partial t^2} \tag{6.2}$$

where $w(x,t)$ is the transverse displacement of the cable, $f(x,t)$ is the lateral force per unit length of the cable, $m(x)$ is the mass per unit length of the cable, T and θ are the cable tension and cable slope at location x, respectively.

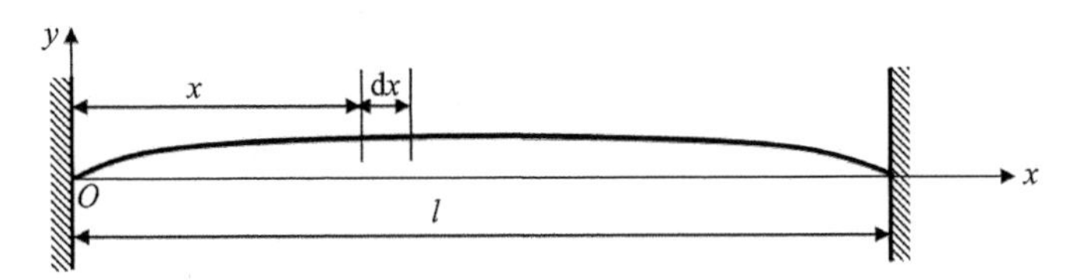

(a) Transverse vibration of a cable in tension

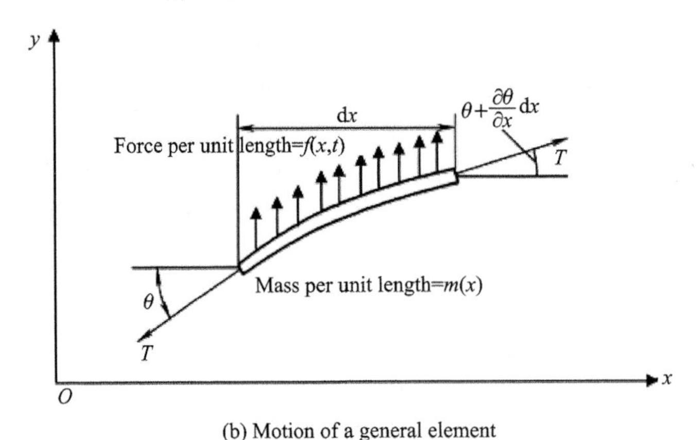

(b) Motion of a general element

Fig. 6.1 Vibration of the cable

Using the small slope assumption in Eq. (6.1), so $\sin(\theta+\mathrm{d}\theta) = \theta + \mathrm{d}\theta$. Eq. (6.2) can be rewritten as

$$m(x)\,\frac{\partial^2 w(x,t)}{\partial t^2} = T\,\frac{\partial^2 w(x,t)}{\partial x^2} + f(x,t)\mathrm{d}x \tag{6.3}$$

We consider the case of free vibration, it means that $f(x,t) = 0$. So Eq. (6.3) can be simplified as the wave equation:

$$\frac{\partial^2 w(x,t)}{\partial t^2}=\frac{T}{m}\frac{\partial^2 w(x,t)}{\partial x^2}=c^2\frac{\partial^2 w(x,t)}{\partial x^2} \tag{6.4}$$

where $c=\sqrt{\frac{T}{m}}$, is termed as wave speed. In other words, c is the speed propagation of waves in the cable.

Assume that the solution of Eq. (6.4) can be expressed as

$$w(x,t)=Y(x)q(t) \tag{6.5}$$

From Eq. (6.5), it can be found that, the structure will have a shape given by $Y(x)$ at any given time t. It means that the structure will maintain a particular "shape" $Y(x)$ at all times. So $Y(x)$ will be a **mode shape.**

Also, at a given pointx of the structure, the structure will vibrate according to the time response $q(t)$. It will be shown that $q(t)$ will obey the simple harmonic motion of a specific frequency. This is the natural frequency of vibration corresponding to that particular mode.

It should be noticed that, for a continuous system, there will be an infinite degree of freedom. So Eq. (6.4) has infinite number of solutions with different natural frequencies. The corresponding functions $Y(x)$ will be orthogonal each other. And the systems will be able to move independently in each mode, and this collection of solutions of the form in Eq. (6.5) will be a complete set. With this qualitative understanding, now we can seek a solution of the form of equation in Eq. (6.5) for the system in Eq. (6.4). Substituting Eq. (6.5) into Eq. (6.4), we get

$$Y(x)\frac{\partial^2 q(t)}{\partial t^2}=c^2\frac{\partial^2 Y(x)}{\partial x^2}q(t) \tag{6.6}$$

Eq. (6.6) can be rewritten as

$$\frac{1}{c^2 q(t)}\frac{\partial^2 q(t)}{\partial t^2}=\frac{1}{Y(x)}\frac{\partial^2 Y(x)}{\partial x^2} \tag{6.7}$$

In Eq. (6.7), it can be found that the left-hand terms are a function of t only, and the right-hand terms are a function of x only. Notice that space x and time t are independent of each other. If the two sides to be equal (see Eq. (6.7)), each function should be a constant. So Eq. (6.7) can be expressed as

$$\frac{1}{c^2 q(t)}\frac{\partial^2 q(t)}{\partial t^2}=s \quad \text{or} \quad \frac{\partial^2 q(t)}{\partial t^2}-sc^2 q(t)=0 \tag{6.8a}$$

$$\frac{1}{Y(x)}\frac{\partial^2 Y(x)}{\partial x^2}=s \quad \text{or} \quad \frac{\partial^2 Y(x)}{\partial x^2}-sY(x)=0 \tag{6.8b}$$

where s is a unknown constant.

From Eq. (6.8a), it can be found that if the constant s is positive, the function $q(t)$ will NOT be oscillatory and transient, which is contrary to the nature of undamped vibration. So it means that s should be negative, we can denot by $s = -\lambda^2$, and Eq. (6.8) can be rewritten as

$$\frac{\partial^2 q(t)}{\partial t^2} + \lambda^2 c^2 q(t) = 0 \tag{6.9a}$$

$$\frac{\partial^2 Y(x)}{\partial x^2} + \lambda^2 Y(x) = 0 \tag{6.9b}$$

From Eq. (6.9b), it is easy to find the solution as

$$Y(x) = A_1 \exp(\mathrm{i}\lambda x) + A_2 \exp(-\mathrm{i}\lambda x) = C_1 \sin(\lambda x) + C_2 \cos(\lambda x) \tag{6.10}$$

Since $Y(x)$ is a real function representing a mode shape, the constants A_1 and A_2 in Eq. (6.10) should be complex conjugates, and C_1 and C_2 should be real. By using Euler's formula, we get

$$\sin(\lambda x) = \frac{\exp(\mathrm{i}\lambda x) - \exp(-\mathrm{i}\lambda x)}{2\mathrm{i}} \quad \text{and} \quad \cos(\lambda x) = \frac{\exp(\mathrm{i}\lambda x) + \exp(-\mathrm{i}\lambda x)}{2} \tag{6.11}$$

It is easy to found that

$$A_1 = \frac{C_2 - \mathrm{i}C_1}{2} \quad \text{and} \quad A_2 = \frac{C_2 + \mathrm{i}C_1}{2} \tag{6.12}$$

One can obtain the complete solution for free vibration of auniform cable that is fixed at both ends. The applicable boundary conditions for fixed cable are

$$Y(0) = Y(l) = 0 \tag{6.13}$$

wherel is the length of the cable. Substituting Eq. (6.13) into Eq. (6.10), we get

$$C_2 = 0 \quad \text{and} \quad C_1 \sin(\lambda l) = 0 \tag{6.14}$$

A possible solutionin Eq. (6.14) is $C_1 = 0$. But this is the trivial solution, which corresponds to $Y(x) = 0$; that is, a stationary cable with no vibration. It follows that the applicable, nontrivial solution is

$$\sin(\lambda l) = 0 \tag{6.15}$$

Eq. (6.15) can produce an **infinite** number of solutions for λ given by

$$\lambda_k = \frac{k\pi}{l} \quad \text{for} \quad k = 1, 2, \cdots, \infty \tag{6.16}$$

Substituting Eqs. (6.14)— (6.16) into Eq. (6.10), the corresponding infinite number of mode shapes can be obtained

$$Y_k(x) = C_k \sin\left(\frac{k\pi}{l} x\right) \tag{6.17}$$

Next step is to determine the corresponding time response (generalized coordinates) $q_i(t)$. Substituting Eq. (6.16) into Eq. (6.9a), we get

$$\frac{\partial^2 q_k(t)}{\partial t^2} + \omega_k^2 q(t) = 0 \tag{6.18a}$$

and

$$\omega_k = \lambda_k c = \frac{k\pi}{l}\sqrt{\frac{T}{m}} \tag{6.18b}$$

Eq. (6.18a) represents a simple harmonic motion with the natural frequencies ω_i given by Eq. (6.18b). It follows that there are an infinite number of natural frequencies, as mentioned earlier. As done in the previous chapters, the general solution of Eq. (6.18a) is given by

$$q_k(t) = B_k \sin(\omega_k t) + D_k \cos(\omega_k t) \tag{6.19}$$

where the parameters B_k and D_k are determined by initial conditions.

Clearly, the general free response of the cable can be expressed as a linear superposition of all modes.

$$w(x,t) = \sum_{k=1}^{\infty} Y_k(x) q_k(t) = \sum_{k=1}^{\infty} [B_k \sin(\omega_k t) + D_k \cos(\omega_k t)] \sin\left(\frac{k\pi}{l}x\right) \tag{6.20}$$

In the discrete case, the eigenvectors can be scaled arbitrarily. Similarly, the eigenfunctions $Y_k(x)$ of the continuous system are determined up to a scalar constant C_k, as shown in Eq. (6.17). $C_k = 1$ is used in Eq. (6.20).

We know that resonance occurs when a MDOF system is excited by a harmonic force with frequency that is equal to a natural frequency. Similarly, the cable vibrates in resonance when a harmonic force $F_0 \sin(\omega_k t)$ is applied to the cable, where F_0 is a constant and ω_k is one of the natural frequencies (as shown in Eq. (6.18b)) of the cable.

From Eq. (6.20), the complete solution for cable has been expressed as a summation of the modal solutions. This is known as the modal series expansion. Such a solution is quite justified because of the fact that the mode shapes are orthogonal each other, and what was obtained above is a complete set of normal modes. The system is able to move independently in each mode, with a unique spatial shape, at the corresponding natural frequency, because each modal solution is separable into a space function $Y_k(x)$ and a time function $q_k(t)$.

Assume that the initial conditions (at $t=0$) for Eq. (6.20) are

$$\left.\begin{aligned} w(x,0) &= f(x) \\ \frac{\partial w(x,0)}{\partial t} &= g(x) \end{aligned}\right\} \tag{6.21}$$

where $f(x)$ and $g(x)$ describe the displacement and velocity of the cable at the time $t=0$, respectively.

Substituting Eq. (6.20) into Eq. (6.21), yields

$$\left.\begin{aligned} w(x,0) &= \sum_{k=1}^{\infty} D_k \sin\left(\frac{k\pi}{l}x\right) = f(x) \\ \frac{\partial w(x,0)}{\partial t} &= \sum_{k=1}^{\infty} B_k \omega_k \sin\left(\frac{k\pi}{l}x\right) = g(x) \end{aligned}\right\} \tag{6.22}$$

Recall the Fourier series presented in Eq. (4.37),

$$F(t) = \frac{a_0}{2} + \sum_{k=1}^{\infty} \left[a_k \cos(k\omega t) + b_k \sin(k\omega t)\right] \tag{6.23}$$

where $a_k = \frac{2}{T}\int_0^T F(t)\cos(k\omega t)\mathrm{d}t$ and $b_k = \frac{2}{T}\int_0^T F(t)\sin(k\omega t)\mathrm{d}t$.

Compare Eq. (6.22) and Eq. (6.23), it can be found that

$$\left.\begin{aligned} B_k &= \frac{2}{l\omega_k}\int_0^l g(t)\sin\left(\frac{k\pi x}{l}\right)\mathrm{d}x \\ D_k &= \frac{2}{l}\int_0^l f(t)\sin\left(\frac{k\pi x}{l}\right)\mathrm{d}x \end{aligned}\right\} \tag{6.24}$$

Example 6.1

A cable of length l shown in Fig. 6.2 is fixed at both ends. If it is plucked at $x = 6/l$, released the cable with zero velocity. Determine the free vibration of the cable.

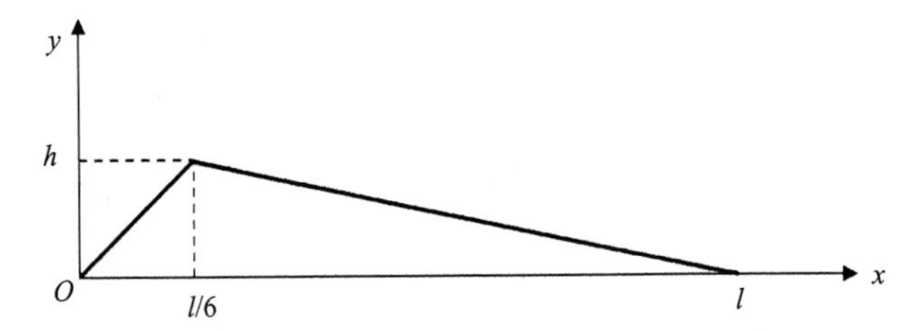

Fig. 6.2 A cable model

Solution:

From Fig. 6.2, it can be found that the initial conditions are

$$w(x,0) = \begin{cases} \frac{6h}{l}x & 0 \leqslant x \leqslant \frac{6}{l} \\ \frac{6h}{5l}(l-x) & \frac{6}{l} \leqslant x \leqslant l \end{cases} \tag{6.25a}$$

and

$$\frac{\partial w(x,0)}{\partial t} = 0 \tag{6.25b}$$

Substituting Eq. (6.25) into Eq. (6.24), we get

$$B_k = 0 \tag{6.26a}$$

$$\left.\begin{aligned} D_k &= \frac{2}{l}\int_0^{\frac{l}{6}} \frac{6hx}{l}\sin\left(\frac{k\pi x}{l}\right)\mathrm{d}x + \frac{2}{l}\int_{\frac{l}{6}}^{l} \frac{6h(l-x)}{5l}\sin\left(\frac{k\pi x}{l}\right)\mathrm{d}x \\ &= \frac{12h}{l^2}\int_0^{\frac{l}{6}} x\sin\left(\frac{k\pi x}{l}\right)\mathrm{d}x + \frac{12h}{5l^2}\int_{\frac{l}{6}}^{l}(l-x)\sin\left(\frac{k\pi x}{l}\right)\mathrm{d}x \\ &= \frac{72h}{5(k\pi)^2}\sin\left(\frac{k\pi}{6}\right) \end{aligned}\right\} \quad (6.26\text{b})$$

Substituting Eq. (6.26) into Eq. (6.20), the solution for this case can be obtained as

$$w(x,t) = \frac{72h}{5\pi^2}\sum_{k=1}^{\infty}\left[\frac{1}{k^2}\sin\left(\frac{k\pi}{6}\right)\cos(\omega_k t)\right]\sin\left(\frac{k\pi}{l}x\right) \quad \text{with} \quad \omega_k = \frac{k\pi}{l}\sqrt{\frac{T}{m}} \tag{6.27}$$

6.2 Transverse Vibration of Beams

6.2.1 The governing equation of motion for beams

This section will discuss yet another continuous member in vibration. Specifically, a beam (or rod or shaft) in flexural vibration is considered. The vibration is in the "transverse" or "lateral" direction. It should be noticed that a beam can support shear forces and bending moments at its cross section, it is unlike to a cable. In the analysis of bending vibration, assume that there is no axial force at the ends of the beam. Further simplifying assumptions will be made, which will be clear in the development of the governing equation of motion. The analysis procedure will be quite similar to that followed in the previous sections for cables.

The study of bending vibration (or lateral or transverse vibration) of beams is very important in avariety of practical situations, such as for bridges, vehicles, tall buildings, aircraft and space stations.

Consider a beam in bending, in the x-y plane, with x as the longitudinal axis and y as the transverse axis of bending deflection, as shown in Fig. 6.3. The required equation is developed by considering the bending moment-deflection relation, rotational equilibrium, and transverse dynamics of a beam element.

1. Moment-deflection relation

A small beam element of length δx subjected to bending moment M is shown in Fig. 6.3. Neglect any transverse deflections due to shear stresses. Consider a strip-like area element δA in the cross section A of the beam element, at a distance w (measure parallel to y) from the neutral axis of bending. The normal strain ε can be expressed as

$$\varepsilon = \frac{(R + w)\delta\theta - R\delta\theta}{R\delta\theta} \tag{6.28}$$

Notice that the neutral axis joins the points along the beam where the normal strain and stressare zero.

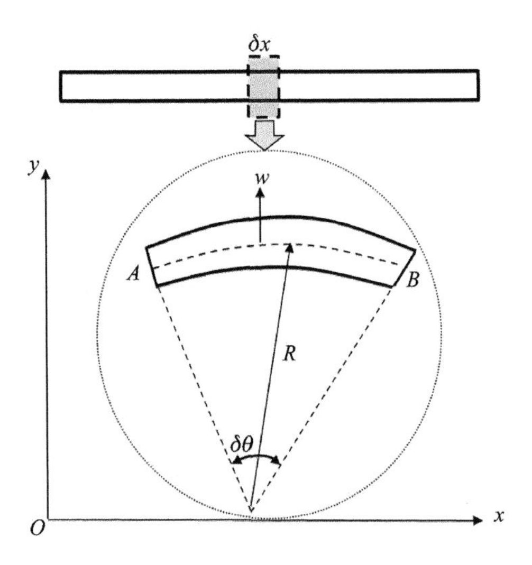

Fig. 6.3 A thin beam in bending

Hence

$$\varepsilon = \frac{w}{R} \tag{6.29}$$

where R is the radius of curvature of the bent element. Normal stress in the axial direction is

$$\sigma = E\varepsilon = E\frac{w}{R} \tag{6.30}$$

where E is Young's modulus (of elasticity). Then the bending moment is

$$M = \int_A w\sigma \mathrm{d}A = \frac{E}{R}\int_A w^2 \mathrm{d}A = \frac{EI}{R} \tag{6.31}$$

where I is the second moment of area of the beam cross section, about the neutral axis. Thus, from Fig. 6.3, it can be found that slope at A and B point are $\frac{\partial w}{\partial x}$ and $\frac{\partial w}{\partial x} + \frac{\partial^2 w}{\partial x^2}\delta x$, respectively, and w is the transverse displacement at element δx. So the change in slope is $\delta\theta = \frac{\partial^2 w}{\partial x^2}\delta x$, also notice that $\delta x = R\delta\theta$, so we can obtain

$$\frac{1}{R} = \frac{\partial^2 w}{\partial x^2} \tag{6.32}$$

Substituting Eq. (6.32) into Eq. (6.31), yields

$$M = EI \frac{\partial^2 w}{\partial x^2} \tag{6.33}$$

2. Rotatory dynamics (equilibrium)

Again consider the beam element δx, as shown in Fig. 6.4, where forces and moments acting on the element are indicated. Here, $f(x,t)$ is the excitation force per unit length acting on the beam, in the transverse direction, at location x. Neglect rotatory inertia of the beam element.

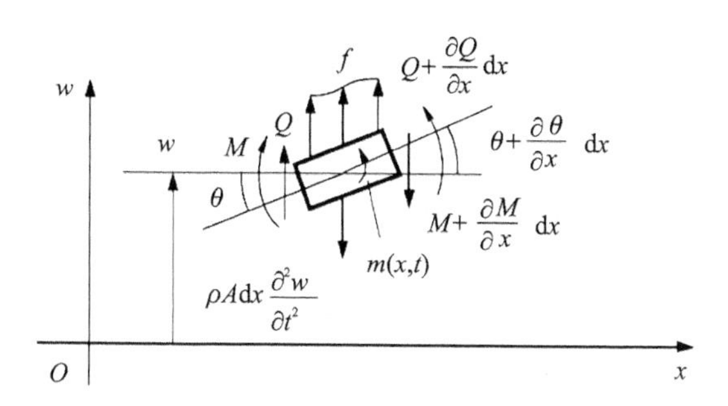

Fig. 6.4 Dynamics of a beam element in bending

The equation of angular motion is given by the equilibrium condition of moments, the shear force Q can be expressed as

$$Q = \frac{\partial M}{\partial x} = \frac{\partial}{\partial x}\left(\frac{\partial^2 w}{\partial x^2}\right) \tag{6.34}$$

3. Transverse dynamics

The equation of transverse motion (Newton's second law) for element δx is

$$(m\delta x)\frac{\partial^2 w}{\partial t^2} = f(x,t)\delta x + Q - \left(Q + \frac{\partial Q}{\partial x}\delta x\right) \tag{6.35}$$

where $m = \rho A$, is the mass per unit length; and ρ is the density of the beam material.

Substituting Eq. (6.34) into Eq. (6.35), the partial differential equation describing the response of a beam under transverse vibration can be obtained

$$\frac{d^2}{dx^2}\left[EI\frac{d^2 w(x,t)}{dx^2}\right] + m\frac{d^2 w(x,t)}{dt^2} = 0 \tag{6.36}$$

6.2.2 The natural frequencies and mode shapes for beams

If the beam is uniform, such as $I(x) = I$, Eq. (6.36) can be rewritten as

$$\frac{d^4 w(x,t)}{dx^4} + \frac{m}{EI}\frac{d^2 w(x,t)}{dt^2} = 0 \tag{6.37}$$

Similar to Section 6.1, the displacement of the beam can be separable in space and time:

$$w(x,t)=\Phi(x)\eta(t) \tag{6.38}$$

where $\Phi(x)$ and $\eta(t)$ are the structural mode shape and the modal coordinate, respectively.

Substituting Eq. (6.38) into Eq. (6.37) and separating variable for time t and space x, two ordinary differential equations are obtained

$$\frac{\mathrm{d}^4\Phi(x)}{\mathrm{d}x^4}-\frac{m\omega^2}{EI}\Phi(x)=0 \tag{6.39a}$$

$$\frac{\mathrm{d}^2\eta(t)}{\mathrm{d}t^2}+\omega^2\eta(t)=0 \tag{6.39b}$$

where ω is the natural frequency of the beam.

Eq. (6.39a) can be rewritten as

$$\frac{\mathrm{d}^4\Phi(x)}{\mathrm{d}x^4}-k^4\Phi(x)=0 \tag{6.40}$$

and

$$k^4=\frac{m\omega^2}{EI} \tag{6.41}$$

To solve Eq. (6.40), the boundary conditions for beam should be considered. For beam-type structures, the common boundary conditions are as follows.

(1) Simply supported boundary condition

$$w(x)=\frac{\partial^2 w(x)}{\partial x^2}=0 \quad (x=0 \text{ or } L_x) \tag{6.42}$$

(2) Clamped boundary condition

$$w(x)=\frac{\partial w(x)}{\partial x}=0 \quad (x=0 \text{ or } L_x) \tag{6.43}$$

(3) Free boundary condition

$$\frac{\partial^2 w(x)}{\partial x^2}=\frac{\partial^3 w(x)}{\partial x^3}=0 \quad (x=0 \text{ or } L_x) \tag{6.44}$$

The general solution to the differential Eq. (6.40) is

$$\Phi(x)=A\sin(kx)+B\cos(kx)+C\sinh(kx)+D\cosh(kx) \tag{6.45}$$

For simply supported boundary condition, at the left end ($x = 0$), substituting Eq. (6.42) into Eq. (6.45), one obtains

$$\Phi(0)=B+D=0 \tag{6.46}$$

$$\frac{\partial^2\Phi(0)}{\partial x^2}=k^2(-B+D)=0 \tag{6.47}$$

Thus $B = D = 0$.

At the right end ($x = L_x$), we get

$$w(L_x)=A\sin(kL_x)+C\sinh(kL_x)=0 \tag{6.48}$$

$$\frac{\partial^2 w(L_x)}{\partial x^2}=k^2\left[-A\sin(kL_x)+C\sinh(kL_x)\right]=0 \tag{6.49}$$

From Eqs. (6.48) and (6.49), we get

$$A\sin(kL_x)=C\sinh(kL_x)=0 \tag{6.50}$$

Since sinh $(kL_x)\neq 0$ provide $kL_x\neq 0$, and therefore

$$C=0,\quad k=\frac{n\pi}{L_x} \tag{6.51}$$

Substituting Eq. (6.51) into Eq. (6.45), the nth mode shape function for simply supported beam can be obtained

$$\Phi_n(x)=\sin\left(\frac{n\pi}{L_x}x\right) \tag{6.52}$$

Substituting Eq. (6.51) into Eq. (6.41), the corresponding natural frequencies can be written as

$$\omega_{\mathrm{n}}=\sqrt{\frac{EI}{m}}\left(\frac{n\pi}{L_x}\right)^2 \tag{6.53}$$

The mode shapes are orthogonal with respect to the mass and stiffness distribution:

$$\int_0^{L_x} m\Phi_j(x)\Phi_k(x)\mathrm{d}x=\mu_j\delta_{jk} \tag{6.54}$$

$$\int_0^{L_x} EI\frac{\partial^2\Phi_j(x)}{\partial x^2}\frac{\partial^2\Phi_k(x)}{\partial x^2}\mathrm{d}x=\mu_j\omega_j^2\delta_{jk} \tag{6.55}$$

where $\delta_{jk}=\begin{cases}1 & j=k\\ 0 & j\neq k\end{cases}$, is the Kronecker delta index, μ_j is the modal mass of the jth mode. The modal mass corresponding to mode shapes in Eq. (6.52) is $mL_x/2$.

Due to the mode shapes being orthogonal to each other, the response of the beam can be expressed at any arbitrary point as a linear combination of these mode shape functions, such as

$$w(x,t)=\sum_{n=1}^{\infty}\Phi_{\mathrm{n}}(x)\ \eta_{\mathrm{n}}(t) \tag{6.56}$$

For other boundary conditions, the structural mode shapes and natural frequencies are listed in Table 6.1.

Table 6.1 Structural mode shapes and natural frequencies for a uniform beam

Boundary Conditions	Structural Mode Shape Functions	Natural Frequencies $\omega_{\mathrm{n}}=\sqrt{\frac{EI}{m}}k_n^2$
Simply-supported	$\Phi_n(x)=\sqrt{\frac{2}{mL_x}}\sin\left(\frac{n\pi}{L_x}x\right)$	$k_n=\frac{n\pi}{L_x}$

(continued)

Boundary Conditions	Structural Mode Shape Functions	Natural Frequencies $\omega_n = \sqrt{\frac{EI}{m}} k_n^2$
Clamped-clamped	$\Phi_n(x) = \cosh(k_n x) - \cos(k_n x) - \beta_n [\sinh(k_n x) - \sin(k_n x)]$ $\beta_n = \frac{\cosh(k_n L_x) - \cos(k_n L_x)}{\sinh(k_n L_x) - \sin(k_n L_x)}$	$\cos(k_n L_x) \cdot \cosh(k_n L_x) - 1 = 0$
Clamped-free	$\Phi_n(x) = \cosh(k_n x) - \cos(k_n x) - \beta_n [\sinh(k_n x) - \sin(k_n x)]$ $\beta_n = \frac{\cosh(k_n L_x) + \cos(k_n L_x)}{\sinh(k_n L_x) + \sin(k_n L_x)}$	$\cos(k_n L_x) \cdot \cosh(k_n L_x) + 1 = 0$
Clamped-simply supported	$\Phi_n(x) = \cosh(k_n x) - \cos(k_n x) - \beta_n [\sinh(k_n x) - \sin(k_n x)]$ $\beta_n = \frac{\cosh(k_n L_x) - \cos(k_n L_x)}{\sinh(k_n L_x) - \sin(k_n L_x)}$	$\tan(k_n L_x) - \tanh(k_n L_x) = 0$

6.2.3 MATLAB examples

Consider equations in Table 6. 1 for this example. Below we give the MATLAB program, which is used to calculate the first four mode shapes for beam structure. A function show_fig. m is used to show the results.

Line	MATLAB Codes	Comments
1	clear all, close all	Remove all variables in workspace and close all figures. It is a good way to start a new MATLAB script (not function)
2	M=4;	The highest order of mode index
3	KK=linspace(0,1,1e3);	KK = x/L_x is the dimensionless variable. It generatesa row vector KK of 1 000 points linearly spaced form 0 to 1
4	% Simply-supported-simply supported	Short comment in MATLAB file. It is a good way to insert some comment in M-file
5	for m=1:M	Notice: MATLAB is case sensitive, meaning that MATLAB distinguishes between variables with upper-and lower-case names
6	B_ss(:,m)=sin(m * pi * KK);	
7	end	This routine calculates the mode shapes for simply supported beam
8	show_fig(KK, B_ss)	Call function show_fig to show the mode shapes

(continued)

9	% Clamped-clamped	
10	f='cosh(x) * cos(x)-1';	Define a function with an unknown variable x
11	for j=1:M	
12	a1(j)=fzero(f,(j+1/2) * pi);	Find a zero of function f near (j+1/2) * pi, clearly, a1 is the solution of equation f=0
13	b1(j)=(sinh(a1(j))+sin(a1(j)))/(cosh(a1(j))-cos(a1(j)));	Calculate βn
14	end	
15	for m=1:M	This routine calculates the mode shapes for clamped-clamped beam
16	B_cc(:,m)=cosh(a1(m) * KK)-cos(a1(m) * KK)-b1(m) * (sinh(a1(m) * KK)-sin(a1(m) * KK));	
17	end	
18	show_fig(KK, B_cc)	Call function show_fig to show the mode shapes
19	% Clamped-free	Calculate and show the mode shapes for clamped-free beam
20	f='cosh(x) * cos(x)+1';	
21	for j=1:M	
22	a1(j)=fzero(f,(j-1/2) * pi);	
23	b1(j)=(sinh(a1(j))-sin(a1(j)))/(cosh(a1(j))+cos(a1(j)));	
24	end	
25	for m=1:M	
26	B_cf(:,m)=(cosh(a1(m) * KK)-cos(a1(m) * KK))-b1(m) * (sinh(a1(m) * KK)-sin(a1(m) * KK));	
27	end	
28	show_fig(KK, B_cf)	
29	% Clamped-simply supported	Calculate and show the mode shapes for clamped-simply supported beam
30	f='tan(x)-tanh(x)';	
31	for j=1:M	
32	a1(j)=fzero(f,(j+1/4) * pi);	
33	b1(j)=(cosh(a1(j))-cos(a1(j)))/(sinh(a1(j))-sin(a1(j)));	
34	end	

(continued)

35	for m=1:M	
36	B_cs(:,m)=(cosh(a1(m) * KK)−cos(a1(m) * KK))−b1(m) * (sinh(a1(m) * KK)−sin(a1(m) * KK));	Calculate and show the mode shapes for clamped-simply supported beam
37	end	
	show_fig(KK, B_cs)	

Below we give the function show_fig. m, which is used to show the first four mode shapes for beam structure. The first line of a function is the function declaration line and begins with the word function by the argument(s), equality sign, name of the function, and input argument(s), as illustrated in the below example.

Line	MATLAB Codes	Comments
1	function show_fig(KK, B_ss)	Notice: The name of a function, as defined in the first line of the M-file, should be the same as the name of the file without the . m extension
2	y_max=max(max(B_ss)) * 1.1;	
3	figure	Create a new figure
4	plot(KK, B_ss(:,1),'k','linewidth',2);	Plot the 1st mode shape by soild line (black color and linewidth=2 points)
5	hold on	Retain the current plot and certain axes properties so that subsequent graphing commands add to the existing graph, it is quite useful for multiple-line plot
6	plot(KK, B_ss(:,2),'k:','linewidth',2);	Plot the 2nd mode shape by dotted line (black color and linewidth=2 points)
7	plot(KK, B_ss(:,3),'k-.','linewidth',2);	Plot the 3rd mode shape by dash-dot line (black color and linewidth=2 points)
8	plot(KK, B_ss(:,4),'k--','linewidth',2);	Plot the 4th mode shape by dashed line (black color and linewidth=2 points)
9	xlim([0 1]),ylim([-y_max y_max])	Set x-axis and y-axis limits
10	xlabel('\itx/L'), ylabel('Mode shape')	Label the x-and y-axis. "\it" is for italic font
11	legend('1st mode','2nd mode','3rd mode','4th mode',0)	Display a legend on the figure. "0" is to determine to place the legend on the best location

After run this MATLAB program, the structural mode shape functions of different boundary conditions listed in Table 6.1 are obtained and shown in Figs. 6.5—6.8.

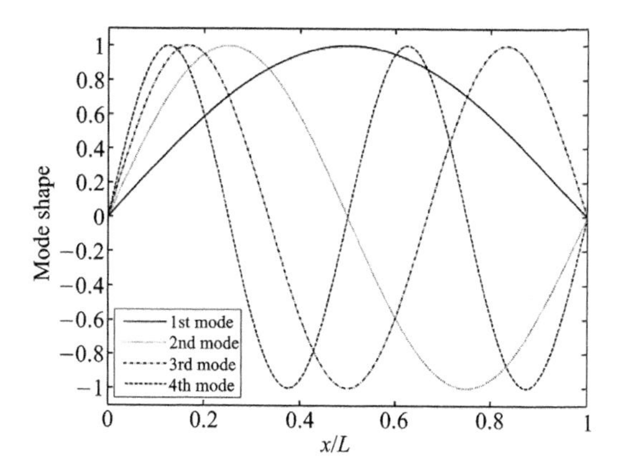

Fig. 6.5 The first four structural mode shapes for simply supported beam

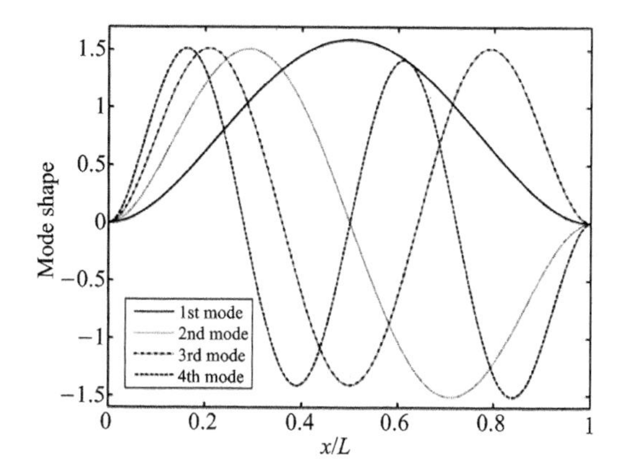

Fig. 6.6 The first four structural mode shapes for clamped-clamped beam

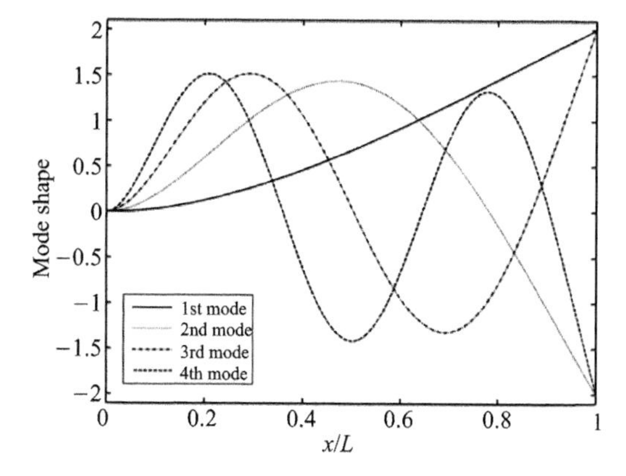

Fig. 6.7 The first four structural mode shapes for clamped-free beam

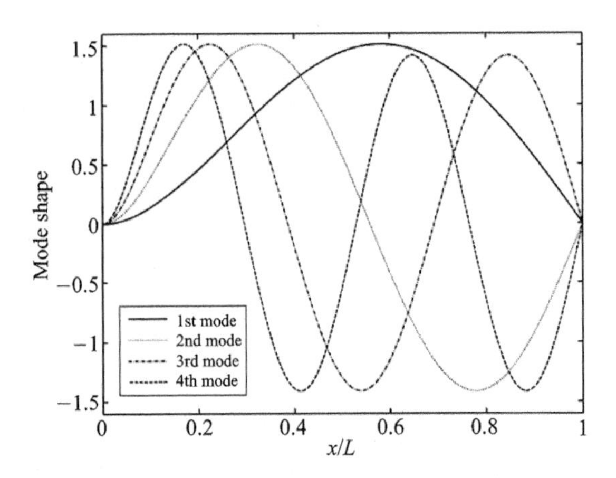

Fig. 6.8 The first four structural mode shapes for clamped-simply supported beam

6.3 The Structural Modes for Plates

The previous section discussed the vibration of beam-type structures. In this section, we will extend the two-dimensional structure vibration. The governing equation for free vibration of an isotropic, non-damping plate can be written as

$$D\nabla^4 w(x,y,t)+m_s\frac{\partial^2 w(x,y,t)}{\partial t^2}=0 \tag{6.57}$$

$$\nabla^2=\frac{\partial^2}{\partial x^2}+\frac{\partial^2}{\partial y^2} \tag{6.58}$$

$$D=\frac{(h)^3 E}{12(1-\nu^2)} \tag{6.59}$$

where $w(x, y, t)$ is the transverse displacement, E is Yong's modulus, $m_s=\rho h$ and ρ and h are density and thickness of the plate, ν is Poisson's ratio.

For harmonic free vibration, the $w(x, y, t)$ can be expressed as the superposition of an infinite number of mode shape functions $\Phi(x, y)$:

$$w(x,y,t)=\sum_{m=1}^{\infty}\sum_{n=1}^{\infty}\Phi_{mn}(x,y)\eta_{mn}e^{i\omega t} \tag{6.60}$$

with the properties

$$\int_0^{L_x}\int_0^{L_y} m_s\Phi_{mn}(x,y)\Phi_{jk}(x,y)\mathrm{d}y\,\mathrm{d}x=\begin{cases}M_{mn} & m=j,n=k\\ 0 & \text{other}\end{cases} \tag{6.61}$$

where η_{mn} is the modal amplitude of the (m, n) th mode of the plate, M_{mn} is the modal mass.

Substituting Eq. (6.60) into Eq. (6.58), yields

$$D\left(\frac{\partial^4}{\partial x^4}+2\frac{\partial^4}{\partial x^2\partial x^2}+\frac{\partial^4}{\partial y^4}\right)\Phi_{mn}(x,y)-\omega_{mn}^2 m_s\Phi_{mn}(x,y)=0 \tag{6.62}$$

where ω_{mn} is the (m, n)th natural frequency.

The structural mode shape functions can be arbitrarily chosen as long as they are quasi-orthogonal and both of them satisfy the boundary conditions. The mode shape functions can be written as the product of two independent beam functions

$$\Phi_{mn}(x,y)=X_m(x)\cdot Y_n(y) \tag{6.63}$$

The shape functions $X_m(x)$ and $Y_n(y)$ can be arbitrarily chosen if and only if they are quasi-orthogonal and satisfy the boundary conditions. And

$$\int_0^{L_x} X_j(x)X_k(x)\mathrm{d}x=\int_0^{L_x}\frac{\partial^2 X_j(x)}{\partial x^2}\frac{\partial^2 X_k(x)}{\partial x^2}\mathrm{d}x=0 \quad \text{if} \quad j\neq k \tag{6.64}$$

$$\int_0^{L_y} Y_j(y)Y_k(y)\mathrm{d}x=\int_0^{L_y}\frac{\partial^2 Y_j(y)}{\partial y^2}\frac{\partial^2 Y_k(y)}{\partial y^2}\mathrm{d}y=0 \quad \text{if} \quad j\neq k \tag{6.65}$$

From Eq. (6.60), by using the orthogonal relationship in Eqs. (6.64) and (6.65), the natural frequencies are given by

$$\omega_{mn}=\sqrt{\frac{D}{m_s}}\cdot\sqrt{\frac{I_1 I_2+2I_3 I_4+I_5 I_6}{I_2 I_6}} \tag{6.66}$$

and

$$I_1=\int_0^{L_x}\frac{\partial^4 X_m(x)}{\partial x^4}X_m(x)\mathrm{d}x,\quad I_2=\int_0^{L_y}(Y_n(y))^2\mathrm{d}y \tag{6.67a}$$

$$I_3=\int_0^{L_x}\frac{\partial^2 X_m(x)}{\partial x^2}X_m(x)\mathrm{d}x,\quad I_4=\int_0^{L_y}\frac{\partial^2 Y_n(y)}{\partial y^2}Y_n(y)\mathrm{d}y \tag{6.67b}$$

$$I_5=\int_0^{L_y}\frac{\partial^4 Y_n(y)}{\partial y^4}Y_n(y)\mathrm{d}y,\quad I_6=\int_0^{L_x}(X_m(x))^2\mathrm{d}x \tag{6.67c}$$

As for simply supported boundaries, we can choose the shape functions as follows:

$$X_m(x)=\sin\frac{m\pi x}{L_x},\quad Y_n(y)=\sin\frac{n\pi y}{L_y} \tag{6.68}$$

Then substituting Eqs. (6.68) and (6.67) into Eq. (6.66), the natural frequencies can be expressed as

$$\omega_{mn}=\sqrt{\frac{D}{m}}\left[\left(\frac{m\pi}{L_x}\right)^2+\left(\frac{n\pi}{L_y}\right)^2\right] \tag{6.69}$$

Fig. 6.9 shows the result of above program, which calculates the first six structural mode shape functions for a simply supported plate.

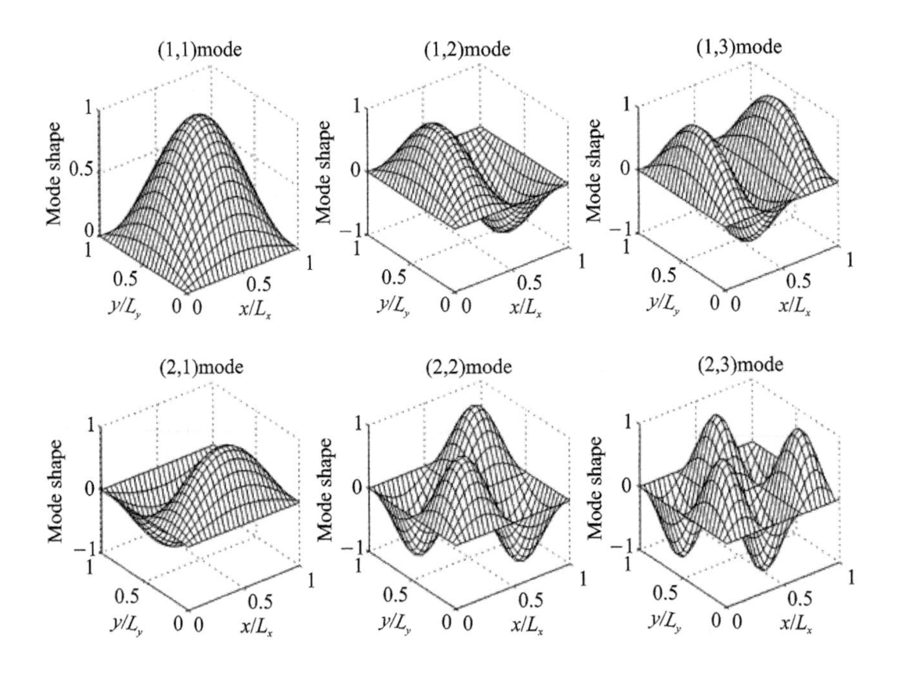

Fig. 6.9 The first six structural mode shapes for a simply supported plate

For a clamped plate, the shape functions can be chosen as follows:

$$X_m(x)=\cosh\frac{\lambda_m x}{L_x}-\cos\frac{\lambda_m x}{L_x}-\beta_m\left(\sinh\frac{\lambda_m x}{L_x}-\sin\frac{\lambda_m x}{L_x}\right) \tag{6.70}$$

$$Y_n(y)=\cosh\frac{\lambda_n y}{L_y}-\cos\frac{\lambda_n y}{L_y}-\beta_n\left(\sinh\frac{\lambda_n y}{L_y}-\sin\frac{\lambda_n y}{L_y}\right) \tag{6.71}$$

where $\beta_m=\dfrac{\cosh\lambda_m-\cos\lambda_m}{\sinh\lambda_m-\sin\lambda_m}$, and λ_m and λ_n are the roots for $\cosh\lambda\cos\lambda=1$.

Notice that $\lambda_i\approx\dfrac{(2i+1)\pi}{2}$ for large values of the integer i.

The natural frequencies for simply supported plates are easy to obtain. It is much more difficult to obtain the natural frequencies for other boundary conditions, i. e. , clamped plate. However, it is possible to solve the issue by using the Symbolic Math Toolbox in MATLAB. This toolbox enables the readers to perform computations using symbolic mathematics and variable-precision arithmetic

Similar to above example, the mode shapes for the clamped plate can be calculated by using Eqs. (6.70) and (6.71), as shown in Fig. 6.10.

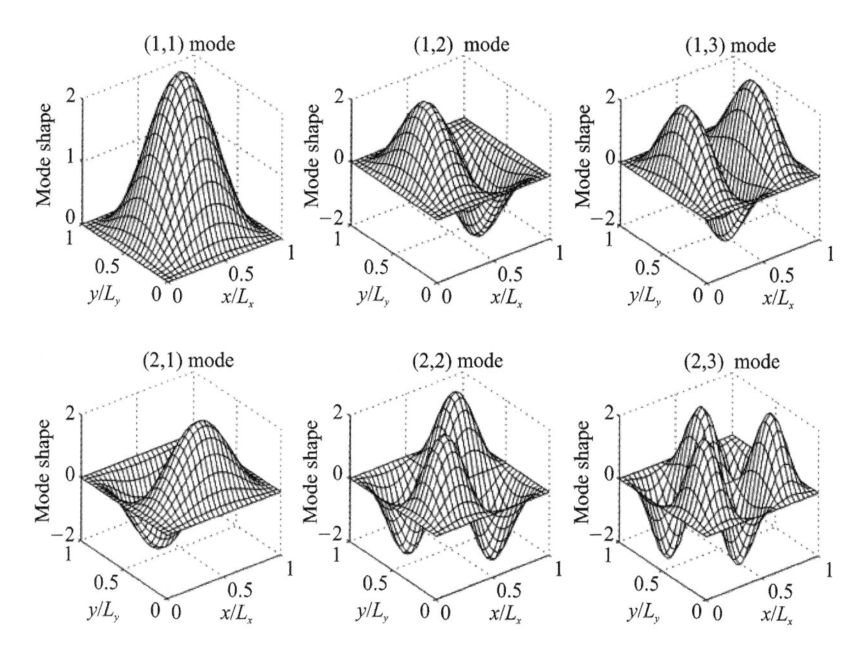

Fig. 6. 10 The first six structural mode shapes for a clamped plate

Questions

6. 1 If a cable of length L, fixed at both ends, is plucked at its midpoint as shown in Fig. 6. 11 and then released, determine its subsequent motion.

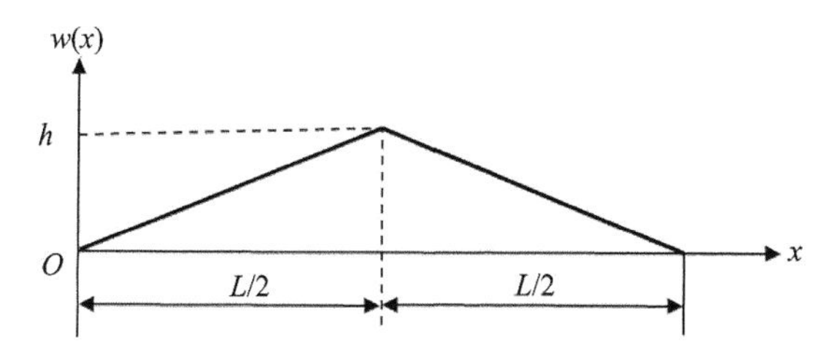

Fig. 6. 11 Initial deflection of the cable

6. 2 Evaluate the first six natural frequencies for the cantilever beam with a mass at the end, as shown in Fig. 6. 12, assuming the end lumped mass $M = mL$ and neglecting its mass moment of inertia. Plot the corresponding structural mode shapes.

6. 3 The uniform beam shown in Fig. 6. 13 is continuous over two spans. Evaluate the first four natural frequencies of this structure and plot the corresponding structural mode shapes.

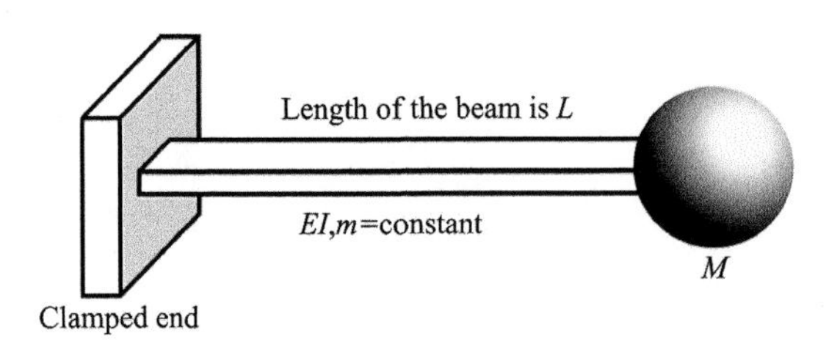

Fig. 6.12　Beam with lumped mass

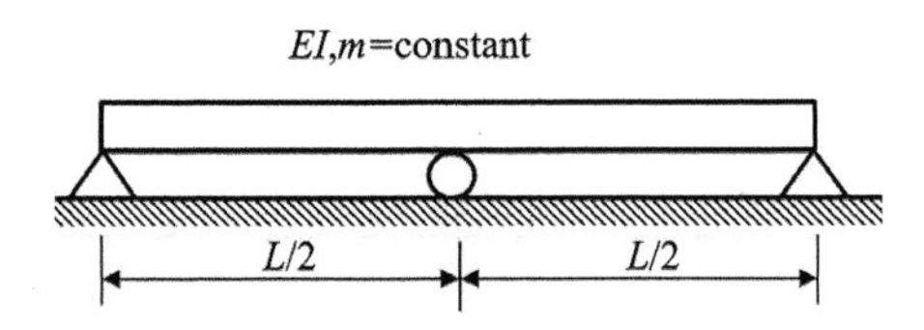

Fig. 6.13　Uniform beam over two spans

6.4　Assume that a thin rectangular plate with length $L_x = 0.5$ m and width $L_y = 0.4$ m, where the short edges are simply supported and the long edges are clamped. Plot the first four structural mode shapes.

Appendix A Mathematical Background

Do not worry about your difficulties in Mathematics. I can assure you mine are still greater.

—Albert Einstein

A. 1 Mathematical Relations

Some of the relationships from trigonometry, algebra, and calculus that are frequently used in vibration analysis are given below

$$\sin(\alpha \pm \beta) = \sin\alpha\cos\beta \pm \cos\alpha\sin\beta$$

$$\cos(\alpha \pm \beta) = \cos\alpha\sin\beta \mp \sin\alpha\cos\beta$$

$$\sin\alpha\sin\beta = \frac{1}{2}[\cos(\alpha - \beta) - \cos(\alpha + \beta)]$$

$$\cos\alpha\cos\beta = \frac{1}{2}[\sin(\alpha + \beta) + \sin(\alpha - \beta)]$$

$$A\sin\alpha + B\cos\alpha = \sqrt{A^2 + B^2}\cos(a - \phi_1)$$

$$= \sqrt{A^2 + B^2}\sin(a + \phi_2)$$

$$\text{with} \quad \phi_1 = \arctan\frac{A}{B} \quad \text{and} \quad \phi_2 = \arctan\frac{B}{A}$$

$$\sin^2\alpha + \cos^2\alpha = 1$$

$$\cos 2\alpha = 1 - 2\sin^2\alpha = 2\cos^2\alpha - 1 = \cos^2\alpha - \sin^2\alpha$$

$$\exp(\mathrm{i}\alpha) = \cos\alpha + \mathrm{i}\sin\alpha \quad \text{with} \quad \mathrm{i} = \sqrt{-1}$$

A. 2 Ordinary Differential Equations (ODE)

The dynamic behavior of mechanical structures and systems can be described by using the second order ODE. The applied forces to the mechanical structure appear on the right hand side of the equation. And the solution of the equation is usually the displacement.

In order to be able to solve these equations, it isnecessary to have a solid background on the solution of homogeneous ODEs. Homogeneous ODEs represent the "free vibrations".

Consider a second order homogeneous linear ODE with constant coefficients a and b.

$$\ddot{y} + a\dot{y} + by = 0 \tag{A. 1}$$

Before solve Eq. (A. 1), we discuss the solution of a first order linear ODE,

$$\dot{y} + ky = 0 \tag{A. 2}$$

By separating variables and integrating, we obtain

$$\frac{\mathrm{d}y}{y} = -k\,\mathrm{d}x \quad \text{and} \quad \ln|y| = -\int k\,\mathrm{d}x + c \tag{A. 3}$$

where c is an unknown constant.

Taking exponents on both sides for Eq. (A. 3), we get

$$y(x) = c\exp\left(-\int k\,\mathrm{d}x\right) = c\exp(-kx) \tag{A. 4}$$

According to the solution given in Eq. (A. 4), and using a constant coefficient λ, we can assume the solution for Eq. (A. 1) as

$$y = \exp(\lambda x) \tag{A. 5}$$

Substituting its derivatives: $\dot{y} = \lambda\exp(\lambda x)$ and $\ddot{y} = \lambda^2\exp(\lambda x)$ into Eq. (A. 1), we get

$$(\lambda^2 + a\lambda + b)\exp(\lambda x) = 0 \tag{A. 6}$$

Hence if λ is a solution of the following important characteristic equation

$$\lambda^2 + a\lambda + b = 0 \tag{A. 7}$$

the exponential solution $y = \exp(\lambda x)$ is a solution of the $\ddot{y} + a\dot{y} + by = 0$.

Now from elementary algebra, it is easy to find that the roots of this quadratic equation are

$$\lambda_1 = \frac{1}{2}(-a + \sqrt{a^2 - 4b}) \tag{A. 8a}$$

$$\lambda_2 = \frac{1}{2}(-a - \sqrt{a^2 - 4b}) \tag{A. 8b}$$

It is easy to find that the functions below are solutions to $\ddot{y} + a\dot{y} + by = 0$, such as

$$y_1 = \exp(\lambda_1 x) \quad \text{and} \quad y_2 = \exp(\lambda_2 x) \tag{A. 9}$$

From algebra we know that the quadratic equation (as shown in Eqs. (A. 7) and (A. 8)) may have three kinds of roots:

Case 1: Two real roots if $a^2 - 4b > 0$

In this case, a basis of solutions of $\ddot{y} + a\dot{y} + by = 0$, in any interval is $y_1 = \exp(\lambda_1 x)$ and $y_2 = \exp(\lambda_2 x)$. Because y_1 and y_2 are defined and real for all x, and y_1/y_2 is not constant. The corresponding general solution can be expressed as

$$y = c_1\exp(\lambda_1 x) + c_2\exp(\lambda_2 x) \tag{A. 10}$$

Case 2: A real double root if $a^2 - 4b = 0$

In this case, there are real double roots $\lambda = -a/2$. If the discriminant $a^2 - 4b$

is zero, we get only one root from Eq. (A. 8), such as

$$\lambda = \lambda_1 = \lambda_2 = -\frac{a}{2} \tag{A. 11}$$

Hence Eq. (A. 7) has only one solution,

$$y_1 = \exp\left(-\frac{a}{2}x\right) \tag{A. 12}$$

To obtain a second independent solution y_2 needed for a basis, we use the method of order of reduction, setting

$$y_2 = uy_1 \tag{A. 13}$$

From Eq. (A. 13), it is easy to obtain that

$$\dot{y}_2 = \dot{u}y_1 + u\dot{y}_1 \tag{A. 14}$$

$$\ddot{y}_2 = \ddot{u}y_1 + 2\dot{u}\dot{y}_1 + u\ddot{y}_1 \tag{A. 15}$$

Substituting Eq. (A. 13) and its derivatives (see Eqs. (A. 14) and (A. 15)) into Eq. (A. 1), we have

$$(\ddot{u}y_1 + 2\dot{u}\dot{y}_1 + u\ddot{y}_1) + a(\dot{u}y_1 + u\dot{y}_1) + buy_1 = 0 \tag{A. 16}$$

By collecting terms, Eq. (A. 16) can be rewritten as

$$\ddot{u}y_1 + \dot{u}(2\dot{y}_1 + ay_1) + u(\ddot{y}_1 + a\dot{y}_1 + by_1) = 0 \tag{A. 17}$$

It is easy to find that theexpression in the last parenthesis in Eq. (A. 17) is zero, since y_1 is a solution of $\ddot{y} + a\dot{y} + by = 0$. The second parenthesis in Eq. (A. 17) is also zero, since $2\dot{y} = -a\exp(-ax/2) = -ay$ (from Eq. (A. 12)). So Eq. (A. 17) can be simplified as

$$\ddot{u}y_1 = 0 \tag{A. 18}$$

From Eq. (A. 18), it is easy to find that $\ddot{u} = 0$, because y_1 is usually nonzero. By integrations twice, we have

$$u = c_1 + c_2 x \tag{A. 19}$$

To get a second independent solution $y_2 = uy_1$ in Eq. (A. 13), we can simply choose $c_1 = 0$ and $c_2 = 1$ and take $u = x$, then Eq. (A. 13) is rewritten as $y_2 = xy_1$. Since these solutions are not proportional, they form a basis. Hence in the case of a double root of $\lambda^2 + a\lambda + b = 0$ a basis of solution of $\ddot{y} + a\dot{y} + by = 0$ on any interval are $\exp\left(-\frac{a}{2}x\right)$ and $x\exp\left(-\frac{a}{2}x\right)$.

The corresponding general solution is

$$y = (c_1 + c_2 x)\exp\left(-\frac{a}{2}x\right) \tag{A. 20}$$

Case 3: Complex conjugate roots if $a^2 - 4b < 0$

In this case, there are two complex roots $-a/2 + \mathrm{i}\omega$ and $-a/2 - \mathrm{i}\omega$. This case occurs if the discriminant of the characteristic equation $\lambda^2 + a\lambda + b = 0$ is negative. In this case, the roots of Eq. (A. 7) are complex, it means that the solutions of the ODE in Eq. (A. 1) could also be complex. However, it is possible to obtain a basis of real solutions for equation $\ddot{y}+a\dot{y}+by=0$, if we assume the y_1 and y_2 as

$$y_1=\exp\left(-\frac{a}{2}x\right)\cos(\omega x),\quad y_2=\exp\left(-\frac{a}{2}x\right)\sin(\omega x)\quad \text{and}\quad \omega^2=b-\frac{1}{4}a^2 \tag{A. 21}$$

It can be verified by substitution Eq. (A. 21) into Eq. (A. 1). They form a basis on any interval because $y_1/y_2=\cot(\omega x)$ is not constant. Hence, a real general solution in Case 3 is

$$y_1=\exp\left(-\frac{a}{2}x\right)[A\cos(\omega x)+B\sin(\omega x)]\quad (A,\ B\ \text{arbitrary}) \tag{A. 22}$$

In summary, the solutions for equation $\ddot{y}+a\dot{y}+by=0$ for different three cases are listed in Table A. 1.

Table A. 1 The solutions for equation $\ddot{y}+a\dot{y}+by=0$ for different three cases

Case	Roots	Basis	General Solutions
1	Distinct real λ_1 and λ_2	$\exp(\lambda_1 x)$ and $\exp(\lambda_2 x)$	$y=c_1\exp(\lambda_1 x)+c_2\exp(\lambda_2 x)$
2	Real double root $\lambda=-a/2$	$\exp(-ax/2)$ and $x\exp(-ax/2)$	$y=(c_1+c_2 x)\exp\left(-\frac{a}{2}x\right)$
3	Complex conjugate $-a/2+i\omega$ and $-a/2-i\omega$	$y_1=\exp\left(-\frac{a}{2}x\right)\cos(\omega x)$ $y_2=\exp\left(-\frac{a}{2}x\right)\sin(\omega x)$	$y_1=\exp\left(-\frac{a}{2}x\right)\cdot$ $[A\cos(\omega x)+B\sin(\omega x)]$

Appendix B Basic of MATLAB

Before reading this appendix:

This appendix is prepared for a complete MATLAB beginner. If you are familiar to MATLAB software, you can skip it.

MATLAB (MathWorks, Inc.) is an interactive program for numerical computation and data visualization. MATLAB is a very powerful software package intended for scientific and engineering computation. It is widely used in industry, academia and government organizations worldwide. MATLAB functionality is contained in three broad areas:

(1) Core package: A wide variety of standard mathematical functions: root finding, matrix manipulations etc. Scripiting language (M-files).

(2) Toolboxes: Specialized packages for control design, statistics, image processing, finance, digital signal processing and much more.

(3) Simulink: Graphical environment for simulation of dynamic systems.

As you will appreciate soon, one of the most convenient features of MATLAB is a unified syntax for real and complex numbers, scalars, vectors and matrices. It is used extensively by vibration and control engineers for analysis and design. There are many different toolboxes available which extend the basic functions of MATLAB into different areas. Furthermore, MATLAB is relatively easy to learn, and the code is optimized to be relatively quick when performing matrix operations. Another advantage is that MATLAB may behave like a calculator or as a programming language.

B.1 How to Read this MATLAB Tutorial

In the sections that follow, the MATLAB prompt (≫) will be used to indicate where the commands are entered. Anything you see after this prompt denotes user input (i. e., a command) followed by a carriage return (i. e., the "enter" key). Often, input is followed by output so unless otherwise specified the line(s) that follow a command will denote output (i. e., MATLAB's response to what you typed in). MATLAB is case-sensitive, **which means that "a + B" is not the same as "a + b"**. Different fonts, like the ones you just witnessed, will also be used to simulate the interactive session. This can be seen in the example below.

For example, MATLAB can work as a calculator. If we ask MATLAB to add two numbers, we will get the answer we expect.

```
>> 5 + 5
ans =
10
```

As we will see, MATLAB is much more than a “fancy” calculator. In order to get the most out this tutorial you arestrongly encouraged to try all the commands introduced in each section and work on all the recommended exercises. This usually works best if after reading this guide once, you read it again (and possibly again and again) in front of a computer.

B.2 Making Matrix

MATLAB is a matrix-based computing environment. All of the data that you enter into MATLAB is stored in the form of a matrix or a multidimensional array. Even a single numeric value like100 is stored as a matrix (in this case, a matrix having dimensions 1 - by - 1):

```
>> a=100;
>> whos a
  Name      SizeBytes  Class     Attributes

  a         1x1     8  double
```

MATLAB uses variables that are defined to be matrices. A matrix is a collection of numerical values that are organized into a specific configuration of rows and columns. Here are examples of matrices that could be defined in MATLAB. The simplest way to create a matrix in MATLAB is to use the matrix constructor operator, [ ]. Create a row in the matrix by entering elements within the brackets. Separate each element with a comma or space. For example, to create a one row matrix of five elements, type

```
>> A = [1 2 3 4 5]
```

To start a new row, terminate the current row with a semicolon.

```
>> A = [1 2 3 4; 5 6 7 8; 9 10 11 12]
```

We get output

```
A =

     1     2     3     4
     5     6     7     8
     9    10    11    12
```

Transpose of a matrix using the apostrophe

```
>> B=A'
```

We get

```
B =

     1     5     9
     2     6    10
     3     7    11
     4     8    12
```

Table B. 1 gives some basic matrix functions.

Table B. 1 Basic matrix functions

Functions	Explanations
ones	Create a matrix or array of all ones
zeros	Create a matrix or array of all zeros
eye	Create a matrix with ones on the diagonal and zeros elsewhere
diag	Create a diagonal matrix from a vector
magic	Create a square matrix with rows, columns, and diagonals that add up to the same number
inv	Inverse of a matrix
det	Determinant of a matrix
trace	Summation of diagonal elements of a matrix
eig	Solve for the eigenvalues

B.2.1 The colon operator

The colon operation ":" is understood by MATLAB to perform special and useful operations. If two integer numbers are separated by a colon, MATLAB will generate all of the integers between these two integers.

```
>> a = 1:8
a =
     1     2     3     4     5     6     7     8
```

If three numbers, integer or non-integer, are separated by two colons, the middle number is interpreted to be a "range" and the first and third are interpreted to be "limits". Thus

```
>>b = 0.0 : .2 : 1.0
b =
         0    0.2000    0.4000    0.6000    0.8000    1.0000
```

By default, MATLAB always increments by exactly 1 when creating the sequence, even if the ending value is not an integral distance from the start:

```
>> A = 1:6.3
A =
     1     2     3     4     5     6
```

B.2.2 Matrix manipulations

To reference a particular element in a matrix, specify its row and column number using the following syntax, where A is the matrix variable. Always specify the row first and column second:

```
A(row, column)
```

For example, for a 4 - by - 4 magic square A,

```
>> A = magic(4)
A =
    16     2     3    13
     5    11    10     8
     9     7     6    12
     4    14    15     1
```

You would access the element at row 4, column 2 with

```
>> A(4,2)
ans =
14
```

In MATLAB, you can access a two-dimensional matrix element with two subscripts: the first representing the row index, and the second representing the column index, as shown in Fig. B. 1.

Column →

(1,1)	(1,2)	(1,3)	(1,4)
(2,1)	(2,2)	(2,3)	(2,4)
(3,1)	(3,2)	(3,3)	(3,4)
(4,1)	(4,2)	(4,3)	(4,4)

Row ↓

Fig. B. 1 The two-dimensional matrix element

B.2.3 The end keyword

MATLAB provides the keyword end to designate the last element in a particular dimension of an array. This keyword can be useful in instances where your program doesn't know (or forget) how many rows or columns there are in a matrix. You can replace the expression in the previous example with

```
>> A(1,end)
ans =
     13
```

B.2.4 Transposing a matrix

Transpose A so that the row elements become columns. You can use either the transpose function or the transpose operator (.') to do this:

```
>> B=A.'
B =
    16     5     9     4
     2    11     7    14
     3    10     6    15
    13     8    12     1
```

Matrix addition, subtraction and multiplication

Assume that we have two matrices:

```
>> E = [1 2;3 4]
E =
     1     2
     3     4
>> F = [2 3;4 5]
F =
     2     3
     4     5
```

If you want to multiply E and F,

```
>> E * F
 ans =
      10     13
      22     29
```

Another option for matrix manipulation is that you can multiply the corresponding elements of two matrices using the ". * " operator (the matrices must be the same size to do this).

```
>> E. * F
 ans =
       2      6
      12     20
```

If you have a square matrix, like E, you can also multiply it by itself as many times as you like by raising it to a given power.

```
>> E^3
 ans =
      37     54
      81    118
```

B.2.5 Element-wise operations

The " * " operator is used for general matrix multiplication (remember that the plain numbers or the vectors are special cases of matrices) and the user is responsible for dimensional compatibility:

```
>>v * A
```

Remember that vector multiplication is given by the inner product. You can obtain the inner product using dot. For 3D vectors only, you can calculate the cross product with cross. In some cases, you will need to perform element-wise multiplication between vectors or to apply an inherently scalar operation to all elements of a vector. Then you must use the special syntax forms ". * "" .^" and "./". Here are some examples. Create two vectors and obtain a third one by element-wise multiplication:

```
>>aa=[0:2:10];
>>bb=[5:-1:0];
>>cc=aa. * bb
```

You're responsible for making sure both vectors have the same length, which you can check with guess which command 1. Now let's raise the elements of one vector to a power given by the elements of another.

Finally, find the reciprocal of each element of the result:

```
>>bases=[-1:5];
>>exponents=[1:7];
>>res1=bases.^exponents
>>reciprocals=1./res1
```

B.3 Functions

MATLAB includes many standard functions. In MATLAB sin and pi denote the trigonometric function sine and the constant π.

```
>> fun=sin(pi/4)
fun =
     0.7071
```

To determine the usage of any function, type

```
>> help function-name
```

Example: Verify the variables i, j, cos, exp,log, log10 in MATLAB.

B.4 Graphics and Plotting

MATLAB is very powerful for specialized graphics. The manual is hundreds of pages long, so we will just cover the very basics.

One of MATLAB the most powerful features is the ability to create graphic plots. Here we introduce the elementary ideas for simply presenting a graphic plot of two vectors. One thing we want to do to with functions is plot them. MATLAB is a very good tool for that. For example, to plot f as a function of x, type plot (x, f). A new window will come up with a plot of $f(x)$ as a function of x. MATLAB has many features for plotting. We will now learn a few of them. First of all we want to define our axes. This is very simple to do.

To define the x-axis, type xlabel('x') in the Command window. Now switch to the window with the figure. You will see a label on the x-axis. Let's do the same for the y-axis. Predictably, to do this you need to type ylabel('f(x)'). Now, if we want to put a title on our graph the command it title('Function f'). As you can see, a lot of the commands are quite intuitive.

Example B. 1

Plot the $(\sin x)/x$ in the interval $[\pi/100, 10\pi]$.

```
>>x=pi/100:pi/100:10 * pi;
>>y=sin(x)./x;
>>plot(x,y);
>>grid
```

The result from above prgram is shown in Fig. B. 2.

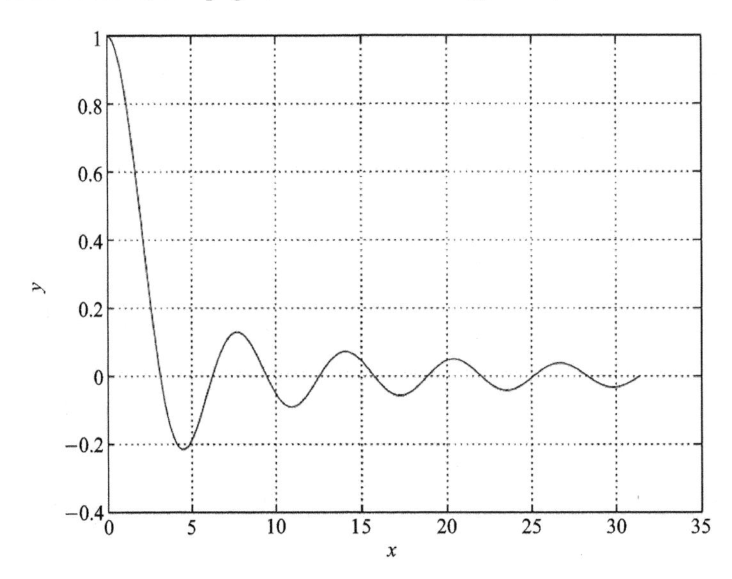

Fig. B. 2 Plot the $(\sin x)/x$

It is also possible to make an animation by MATLAB function drawnow:

```
>>x=linspace(0,10 * 2 * pi,1e3);
>>for m=1:0.05:10
>>plot(x,sin(x * m),'color',[0 0 m/10]),
>>axis([0, 10, −1.2 1.2])
>> text(1,−1.05, ['sin(' num2str(m) ' * x)'])
>>drawnow
>>pause(0.1)
>>end
```

Explanation: You will see a sine function sweeping from 1 rad/s to 10 rad/s, and the line color will be changed from black to blue. The drawnow command flushs event queue and update figure window immediately. Command pause(0. 1) pauses execution for 0. 1 second before continuing, so we can see the animation clearly. Command text adds the string in quotes to the location specified by the point (1, −1.05).

Notice that a new plot replaces previous plots in above example. If you want to add a data set after the plot is done, you can issue the command hold on. Finally, let's do 4 copies of the same 3D plot.

```
>>[X,Y]=meshgrid(-1:0.1:1);
>>Z=X^2-X.*Y+Y^2;
>>subplot(2,2,1)
>>mesh(X,Y,Z)
>>subplot(2,2,2)
>>mesh(X,Y,Z)
>>subplot(2,2,3)
>>mesh(X,Y,Z)
>>subplot(2,2,4)
>>mesh(X,Y,Z)
```

Explanation: Meshgrid creates specially formatted matrices X and Y so that the calculation used for Z takes all possible combinations between elements of X and Y (a grid of points on the X - Y plane, if you will). Note that we must use the ".*" syntax for multiplication.

The first two arguments to subplot are the number of rows and columns for the plot array.

The third indicates the active plot. Numbering is from left to right and from top to bottom, as you can see. Feel free to use the interactive zoom and rotation tools.

Table B.2 gives some basic plotting command

Table B.2 Basic plotting commands

Commands	Explanations
plot(x,y)	A Cartesian plot of the vectors x and y
subplot	Display multiple plots in the same window
loglog	A plot of log(x) vs. log(y)
semilogx(x,y)	A plot of log(x) vs. y
semilogy(x,y)	A plot of x vs. log(y)
title	Placing a title at top of graphics plot
xlabel	Label for x-axis
ylabel	Label for y-axis

(continued)

Commands	Explanations
grid	Creating a grid on the graphics plot
bar	Draw bar graph
pie	DrawPie chart
mesh	Draw a wireframe mesh
surf	Create a three-dimensional shaded surface
image	Display image object

B.5 Programming in MATLAB

B.5.1 The M-files

It is convenient to write a number of lines of MATLAB code before executing the commands. Files that contain a MATLAB code are called the M-files. Use the MATLAB Editor to write code in the MATLAB language and save them for future use. The editor saves your files with the m extension, which is where the name of M-file originates. There are two kinds of M-files: Scripts, which do not accept input arguments or return output arguments. They operate on data in the workspace. Functions, which can accept input arguments and return output arguments. Internal variables are local to the function.

1. Scripts

When you invoke a script, MATLAB simply executes the commands found in the file. Scripts can operate on existing data in the workspace, or they can create new data on which to operate. Although scripts do not return output arguments, any variables that they create remain in the workspace, to be used in subsequent computations. In addition, scripts can produce graphical output using functions like plot.

2. Functions

Functions are M-files that can accept input arguments and return out-put arguments. The names of the M-file and of the function should be the same. Functions operate on variables within their own workspace, separate from the workspace you access at the MATLAB command prompt.

B.5.2 Repeating with "for" loops

Suppose we want to do something several times. You can do this by using a very convenient structure called a loop. A loop loops (hence the name) through some commands several times.

Syntax of the "for" loop is shown below:

```
for n=0:10
x(n+1)=sin(pi * n)
end
```

The "for" loops can be nested, for example,

```
H=zeros(5)
for k=1:5
for l=1:5
H(k,l)=sin(k) * cos(k);
end
end
```

B.5.3 "if" statements

"if" statements use relational or logical operations to determine what steps to perform in the solution of a problem.

The general form of a simple "if" statement is

```
if expression
commands
end
```

In the case of the simple "if" statement, if the logical expression is true, the command is executed. However, if the logical expression is false, the command is bypassed and the program control jumps to the statement that follows the end statement.

The if-else statement allows one to execute one set of statements if logical expression is true and a different set of statements if the logical statement is false. The general form is

```
if expression
command (evaluated if expression is true)
else
command (evaluated if expression is false)
end
```

B.5.4 Writing function subroutines

```
function [mean,stdev] = stat(x)
 n = length(x);
 mean = sum(x) / n;
```

B.6 Saving and Loading

All variables in the workspace can be viewed by commandwhos or who.

To save all variables from the workspace in binary MAT-file

```
save FILENAME
```

An ASCII file is a file containing characters in ASCII format, a format that is independent of "MATLAB" or any other executable program. You can save variables from the workspace in ASCII format with option.

```
save filename. dat variable-ascii
```

To load variables you can use "load" command.

```
load FILENAME
```

To clear variables you can use "clear" command.

```
clear
```

To save all commands and workspace replies in a session, type "diary" when you want to start logging. To stop logging type diary again. A text file will be created in the current directory with all of your commands.

To see the current directory type "pwd". To see the files in the current directory quickly, type "dir". Try it by typing diary now and typing in a few commands. Then turn the diary off and look for the file it created and inspect its contents using, for instance the MATLAB editor. The above only saves commands. Suppose you want to save variables x and y from before (hopefully you haven't cleared them, check with who).

Type "save mydata x y". This will create a file named mydata. mat in the current directory. Now do clear all variables with clear. If you type "who", there should be nothing there. Load your data by typing load mydata and type who again. If you had typed just "save mydata", all variables in the workspace would have been saved.

Finally, suppose you captured some data from an experiment and you want to import it into MATLAB. Go to Excel and create some uniform columns of data (the data must be a rectangular block). In Excel, export the data to a plain text file (tabseparated) named export_data. txt. (actually any extension will do). Place the file in the current MATLAB directory (use pwd to see it). Now type "load-ascii export_data". txt. If you type "who", you should see a variable with the same name as the file name. Type size(export_data) to see how many rows and columns exist. To display the data just invoke its name by typing export_data.

B.7 The Help Menu

It is very important to learn more about a function by using MATLAB help and demonstration examples. There are two ways to get help from MATLAB. First, we simply type help followed by the command name when you know it. For example,

```
>>help eig
>> help svd
```

Another way is to open the help window, as shown in Fig. B. 3.

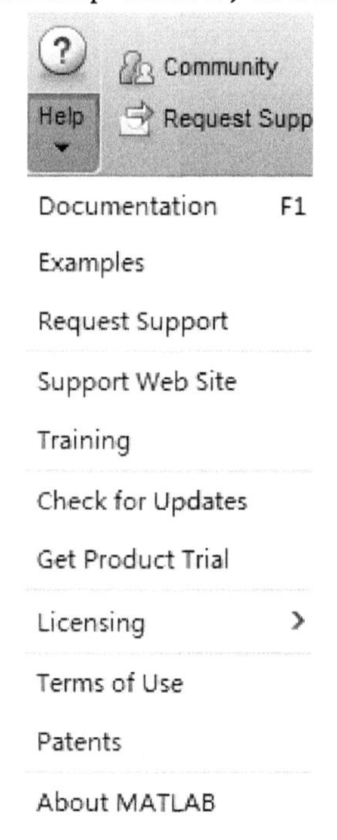

Fig. B. 3 The help menu in MATLAB

In this way, we can get documentation and numerous examples. It is strongly recommended to take a look the MATLAB examples if you are a complete MATLAB beginner.

B. 8 Set the Display Format for Output

In MATLAB, we can use function format to set the display format in workspace. It should be noted that MATLAB always works internally with double precision, but the number of decimals displayed can be adjusted. For example, we can display sin(π/3) in different formats.

```
>> A=sin(pi/3)
A =
      0.8660
>> format shorteng
>> A
A =
      866.0254e-003
>> format long
>> A
A =
      0.866025403784439
>> format longeng
>> A
A =
      866.025403784439e-003
```

B. 9 Closing Remarks and References

It is our hope that by reading this guide you formed a general idea of the capabilities of MATLAB, while obtaining a working knowledge of the program.

There are numerous other commands and specialized functions and toolboxes that may be of interest to you. For example, the MATLAB graphical user interface development environment (GUIDE), provides a set of tools for creating Graphical User Interfaces (GUIs); SIMULINK is a software package for modeling, simulating, and analyzing dynamic systems.

A good source of information related to MATLAB, the creator company The Mathworks, Inc. and their other products is their Web Page at www. mathworks. com. It is strongly recommended that you visit this web page to see what other publications (if any) exist that will allow you to enhance your knowledge of MATLAB.

Appendix C Properties of Laplace Transform and Laplace Transform Pairs

Table C. 1 Properties of Laplace transform

Property	$x(t)$	$X(s)$
Linearity	$a_1x_1(t)+a_2x_2(t)$	$a_1X_1(s)+a_2X_2(s)$
Time shifting	$x(t-a)$	$\exp(-as)X(s)$
Frequency shifting	$\exp(-at)x(t)$	$X(s+a)$
Time differentiation	$\frac{dx}{dt}$	$sX(s)-x(0^+)$
Second derivative	$\frac{d^2x}{dt^2}$	$s^2X(s)-sx(0^+)-x'(0^+)$
Third derivative	$\frac{d^3x}{dt^3}$	$s^3X(s)-s^2x(0^+)-sx'(0^+)-x''(0^+)$
nth derivative	$\frac{d^nx}{dt^n}$	$s^nX(s)-s^{n-1}x(0^+)-s^{n-2}x'(0^+)-\cdots-x^{n-1}(0^+)$
Convolution	$x(t)*h(t)$	$X(s)H(s)$

Table C. 2 Laplace transform pairs (Defined for $t \geqslant 0$, $x(t)=0$ for $t<0$)

No.	$x(t)$	$X(s)$
1	$\delta(t)$	1
2	$s(t)$	$\frac{1}{s}$
3	$s(t)-s(t-a)$	$\frac{1-\exp(as)}{s}$
4	$\exp(-at)$	$\frac{1}{s+a}$
5	t	$\frac{1}{s^2}$
6	t^n	$\frac{n!}{s^{n+1}}$
7	$t\exp(-at)$	$\frac{1}{(s+a)^2}$
8	$t^n\exp(-at)$	$\frac{n!}{(s+a)^{n+1}}$
9	$\sin(\omega t)$	$\frac{\omega}{s^2+\omega^2}$

(continued)

No.	$x(t)$	$X(s)$
10	$\cos(\omega t)$	$\frac{s}{s^2+\omega^2}$
11	$\sin(\omega t+\theta)$	$\frac{s\sin\theta+\omega\cos\theta}{s^2+\omega^2}$
12	$\cos(\omega t+\theta)$	$\frac{s\cos\theta-\omega\sin\theta}{s^2+\omega^2}$
13	$\exp(-at)\sin(\omega t)$	$\frac{\omega}{(s+a)^2+\omega^2}$
14	$\exp(-at)\cos(\omega t)$	$\frac{s+a}{(s+a)^2+\omega^2}$
15	$t\sin(\omega t)$	$\frac{2\omega s}{(s^2+\omega^2)^2}$
16	$t\cos(\omega t)$	$\frac{s^2-\omega^2}{(s^2+\omega^2)^2}$
17	$\sinh(\omega t)$	$\frac{\omega}{s^2-\omega^2}$
18	$\cosh(\omega t)$	$\frac{s}{s^2-\omega^2}$
19	$\frac{1}{\omega_d}\exp(-\zeta\omega_n t)\sin(\omega_d t)$	$\frac{1}{s^2+2\zeta\omega_n s+\omega^2}$
20	$\frac{1-\cos(\omega t)}{\omega^2}$	$\frac{1}{(s^2+\omega^2)s}$

Appendix D Technical Terms

A

absolute motion	绝对运动
acceleration	加速度
accelerometer	加速度计
algebraic	代数的
amplitude	幅值
amplitude-frequency characteristic	幅频特性
angular displacement	角位移
angular velocity	角速度
angular acceleration	角加速度
aperiodic	非周期的
average value	平均值
axis	轴,坐标轴

B

base excitation	基础激励
beam	梁
beating	拍
bending vibration	弯曲振动
boundary condition	边界条件

C

cantilever	悬臂,悬臂梁

Cartesian coordinate	笛卡儿坐标
centrifugal	离心的
centrifugal force	离心力
characteristic determinant	特征行列式
characteristic equation	特征方程
characteristic matrix	特征矩阵
circuit frequency	圆频率
clampedboundary condition	固支边界条件
clockwise	顺时针方向的
coefficient	系数
compress	压缩
concentrated force	集中力,集中载荷
conjugate	共轭的
constant	常数
continuous system	连续系统
continuous vibration system	连续振动系统
converge	收敛
convolution	卷积
convolution integral	卷积积分
column	列
coordinate	坐标
cosine	余弦
counterclockwise	逆时针方向的
coupling	耦合的
critical damping	临界阻尼

cycle	周期

D

damper	阻尼,阻尼器
damped free vibration	阻尼自由振动
damped natural frequency	阻尼固有频率
decelerate	减速
deterministic vibration system	确定振动系统
discrete vibration system	离散振动系统
differential equations of motion	运动微分方程
damping element	阻尼元件
damping	阻尼
damping factor	阻尼系数
damping matrix	阻尼矩阵
damping ratio	阻尼比
datum position	基准位置
decay	衰减
deflection	位移,扰度
denominator	分母
density	密度
derivative	导数
determinant	行列式
diagonal matrix	对角矩阵
differential	微分的
differentiate	微分,对……微分
dimensionless	无量纲的

discrete	离散的
displacement	位移
dissipate	耗散
divide	除,除法
DOF (Degree Of Freedom)	自由度
double pendulum	双摆
Duhamel's integral	杜哈美积分
dynamic	动态的
dynamic response	动态响应
dynamic excitation	动态激励

E

eccentric mass	偏心质量
eccentricity	偏心距
effective mass	有效质量
eigenvalue	特征值
eigenvector	特征向量
element	元素,单元
energy method	能量法
equilibrium	平衡
equilibrium position	平衡位置
exponential	指数的
excitation frequency	激励频率
excitation output	激励输出
elastic element	弹性元件
exciter	激振器

F

FFT(Fast Fourier Transform)	快速傅里叶变换
Finite Element Method (FEM)	有限元法
flexibility	柔度
flexibility influence coefficients	柔度系数
forced vibration	受迫振动
Fourier series	傅里叶级数
frequency shifting	频移
free body diagram	受力分析图
free vibration	自由振动
free vibration of undamped system	无阻尼自由振动
frequency	频率
frequency ratio	频率比

G

general solution	通解
generalized coordinate	广义坐标
generalized force	广义力
gravitational force	重力

H

harmonic excitation	简谐激励
harmonic vibration	简谐振动
harmonic response analysis	谐响应分析
Hertz	赫兹(单位)
homogeneous	齐次的

homogeneous equation	齐次方程
Hooke's law	胡克定律
horizontal	水平的

I

identity matrix	单位矩阵
impulse excitation	冲击激励
impulse response function	冲击响应函数
inertial element	惯性元件
independent coordinate	独立坐标
Industrial Revolution	工业革命
influence coefficient method	影响系数法
initial phase	初始相位
initial condition	初始条件
initial displacement	初始位移
initial velocity	初始速度

J

K

kinetic energy	动能

L

lag	滞后,落后
Lagrange's equation	拉格朗日方程
Lagrangian	拉格朗日算子
Laplace transform	拉普拉斯变换
lead	超前

linear system	线性系统
linearity	线性
linearization	线性化
linear vibration system	线性振动系统
logarithm	对数
logarithmic decrement	对数衰减率
longitudinal vibration	纵向振动
lumped mass	集中质量

M

Magnification Factor (MF)	放大系数,放大因子
magnitude	幅值
mass	质量
mass matrix	质量矩阵
matrix	矩阵
mechanical system	机械系统
mechanical vibration	机械振动
modal analysis	模态分析
modal coordinate	模态坐标
modal superposition	模态叠加
mode shape	模态,振形
moment	力矩
moment of inertia	转动惯量
multi-degree-of-freedom vibration system	多自由度振动系统

N

natural frequency	固有频率
Newton's second law	牛顿第二运动定理
nondimensional	无量纲的
nonlinear vibration system	非线性振动系统
non-damping system	无阻尼系统
normal force	法向力

O

origin	原点,坐标原点
ordinary differential equation	常微分方程
orthogonality	正交
oscillatory	振荡,振荡性
oscillation	振动
overdamped	过阻尼

P

parametric vibration	参数振动
partial differential equation	偏微分方程
particular solution	特解
peak	峰值
period	周期
periodic	周期的
perpetuate	保持,使永存
phase	相位
phase angle	相位角

phase shift	相位差
Poisson's ratio	泊松比
polar coordinate	极坐标
potential energy	势能
principle of superposition	叠加原理
proportional	正比于

Q

quality	质量
quasi-period	准周期

R

random vibration system	随机振动系统
relative damping coefficient	相对阻尼系数
relative motion	相对运动
repetitive	重复的
response output	响应输出
resistance	阻力
resonance	共振
restoring force	回复力
rotary system	旋转系统
rotating unbalance	旋转不平衡
rotational speed	转速

S

self-excited vibration	自激振动
simple harmonic vibration	简谐振动

simple pendulum	单摆
simply supported boundary condition	简支边界条件
single-degree-of-freedom vibration system	单自由度振动系统
sine	正弦
singularity	奇点
slope	斜率,角系数
speed	速度
spring	弹簧
springs connected in series	串联弹簧
springs connected in parallel	并联弹簧
static deflection	静变形
stiffness	刚度
stiffness matrix	刚度矩阵
steady-state response	稳态响应
steady-state solution	稳态解
stretch	伸展,延伸
superposition of vibration	振动叠加
swing	摇摆,摇荡

T

tension	拉力,张力
time domain	时域
time shifting	时移
time differentiation	时域导数
torque	力矩
torsional spring	扭簧

torsional vibration	扭转振动
transfer function	传递函数
transient analysis	瞬态分析
transmissibility of vibration	振动传递
transverse vibration	法向振动,横向振动
trigonometricfunction	三角函数
trivial solution	平凡解
two-degree-of-freedom vibration system	两自由度振动系统

U

undamped system	无阻尼系统
underdamping	欠阻尼
unit matrix	单位矩阵

V

variation	变量,变化
vector	向量
velocity	速度
vibration	振动
vibration mode	振型,振动模态
vibration of elastic body	弹性体振动
vibration system	振动系统
vibration test	振动测试
vibrator	激振器
viscous damping	粘性阻尼

W

wave equation	波动方程
wave speed	波速

X

x-axis	x 轴

Y

y-axis	y 轴

Z

z-axis	z 轴

References

[1] Ni Z. Vibration mechanics [M]. Xi'an: Xi'an Jiaotong University Press, 1989.

[2] Hu H. Basic of mechanical vibration [M]. Beijing: Beihang University Press, 2005.

[3] Liu Y, Chen L, Chen W. Vibration mechanics [M]. 2nd ed. Beijing: Higher Education Press, 2011.

[4] Beards C E. Structural vibration: analysis and damping [M]. London: Butterworth-Heinemann Publisher, 1996.

[5] Rao S S. Mechanical vibrations [M]. 5th ed. Prentice Hall, 2010.

[6] Kreyszig E. Advanced engineering mathematics [M]. 9th ed. New York: Wiley, 2006.

[7] de Silva C W. Vibration: Fundamentals and practice [M]. CRC Press, 1999.

[8] Josephs H, Huston R L. Dynamics of mechanical systems [M]. CRC Press, 2002.

[9] Gatti P L, Ferrari V. Applied structural and mechanical vibrations: Theory, methods and measuring instrumentation [M]. Taylor & Francis Group LLC, 2003.

[10] Gatti P L. Applied structural and mechanical vibrations: Theory and methods [M]. 2nd ed. CRC Press, 2014.

[11] Mao Q, Pietrzko S. Control of noise and structural vibration: a MATLAB-based approach [M]. Springer, 2013.

[12] Shabana A A. Theory of Vibration [M]. Springer, 1991.

[13] Preumont A. Vibration control of active structures: an introduction [M]. 2nd ed. Kluwer, Dordrecht, 2002.

[14] Humar J L. Dynamics of structures [M]. Prentice Hall, Englewood Cliffs, NJ, 1990.

[15] Hahn B, Valentine D T. Essential MATLAB for engineers and scientists [M]. 6th ed. Academic Press, 2016.